风景的情怀

Landscape in my view

王若愚　著

江苏凤凰文艺出版社
JIANGSU PHOENIX LITERATURE AND ART PUBLISHING, LTD

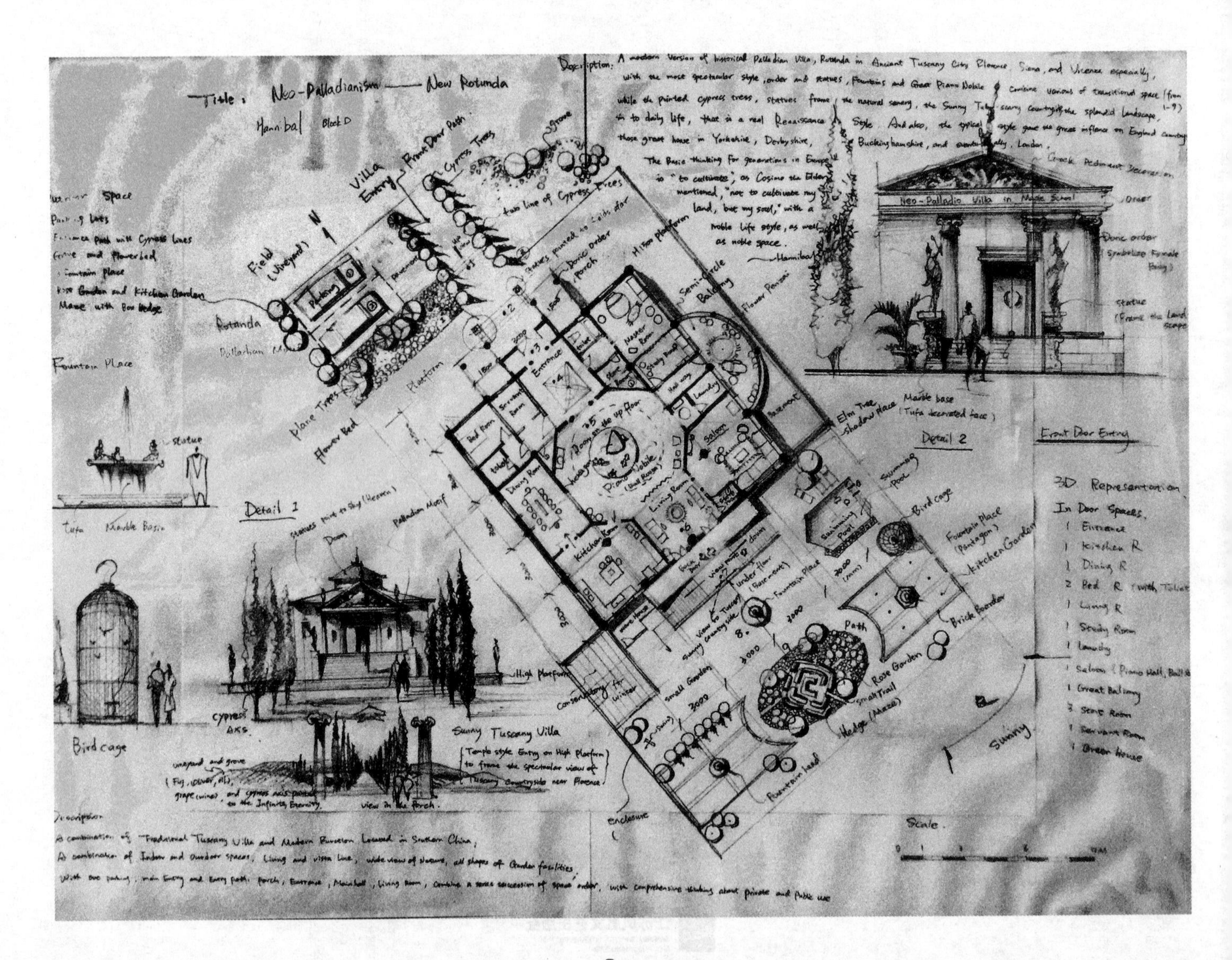
Title: Neo-Palladianism —— New Rotunda
Hannibal Block D
Description: A modern Version of historical Palladian Villa, Rotunda in Ancient Tuscany City Florence, Siena, and Vicenza especially,
with the most spectacular style, order and statues, Fountains and Great Piano Nobile
while the painted cypress trees, statues frame the natural scenery, the Sunny Tuscany countryside, the splendid landscape,
to daily life, that is a real Renaissance Style. And also, the typical style gave the great influence on England country
those great house in Yorkshire, Derbyshire, Buckinghamshire, and eventually, London.
The Basic thinking for generations in Europe is "to cultivate", as Cosimo the Elder mentioned, "not to cultivate my land, but my soul," with a noble life style, as well as noble space.
—Hannibal
Greek Pediment Decoration
Neo-Palladio Villa in Music School
Order
Doric order
Statue
Front Door Entry
Villa Entry
Front Door Path
Cypress Trees
two line of Cypress Trees
Field (Vineyard)
Parking
Rotunda
Fountain Place
Statue
Tufa
Marble Basin
plane Trees
Platform
Flower Bed
Entrance
Living Room
Kitchen Room
Dining Room
Bed Room
toilet
Saloon
Laundry
Study Room
Semi-Circle Balcony
Flower Penzai
Doric Order Porch
Elm Tree shadow place
Marble base (Tufa decorated face)
Detail 1
Detail 2
Palladian Motif
Swimming Pool
Bird cage
Fountain Place (Pentagon)
Kitchen Garden
Brick Border
Path
Rose Garden
Small Trail
Hedge (Maze)
Fountain Place
View to Tuscany Country side
Small Garden
Conservatory for Winter
Fountain head
Enclosure
Sunny
Scale.
3D Representation
In Door Spaces
1 Entrance
1 Kitchen R
1 Dining R
2 Bed R (with Toilet)
1 Living R
1 Study Room
1 Laundry
1 Great Balcony
1 Servant Room
1 Green House
Detail 1
Dome
High Platform
Cypress Axis
Birdcage
Sunny Tuscany Villa
Temple style Entry on High Platform to frame the spectacular view of Tuscany Countryside near Florence.
view in the Porch.
Description
A combination of Traditional Tuscany Villa and Modern Function Located in Southern China,
A combination of Indoor and Outdoor spaces, Living and vista line, wide view of Nature, all shapes of Garden facilities,
With one parking, main Entry and Entry path, Porch, Entrance, Mainhall, Living Room, Combine a series succession of space order, with comprehensive thinking about private and public use

风景的情怀

Landscape in my view

我记不清是什么时候喜欢上画画的，只知道在我很小的时候，家里就有一面空白的墙，是我和父亲涂鸦的地方。从江南小巷的老家，到北京林业大学狭小的宿舍，在墙上涂鸦的这个习惯一直保持着。父亲总会在墙上贴满他的工作草图，偶尔也会在设计之余，贴上几张自己的速写，而这些小幅的草图和速写或许是我学画入门的主要因素。不过当时我很少愿意以这些建筑和风景为绘画题材。属于我的那块白墙，总是会被我涂满各式各样的卡通人物。我的风景园林之旅实际上是从我涂鸦和画漫画人物开始的。二年级的时候我完成了第一个“作品”，美国历史上 30 多位总统的卡通漫画集。这本小册子曾经让从远方回来的父亲大为惊讶，因为他总是觉得画人物比画风景更难以把握，这一点我与父亲的感受不同。也就从那时候开始，父亲随手留在家里的风景速写和涂鸦作品，成了我心摹手追的对象。

在同龄的 00 后人中，很少有人会像我这样经历频繁的搬迁。我的小学、初中乃至高中生活，似乎一直在不停地迁居中度过。从家乡的童年生活，到北京读小学、初中的生活，然后到西安读初中，再到江苏读高中，直到今天来加拿大不列颠哥伦比亚读高中，我的少年时代似乎一直处于不停转学和旅行之中，这种经历除了必然的不利因素以外，却也有一些好处。首先，它使我从很小的时候就习惯随遇而安，并逐步认可、广泛接纳各种形式的学习和生活方式。比如在小学的美术课上，我曾画过一只六条腿的兔子，（具有现代绘画风格中所强调的运动感）这张作业最后被当时给我们讲课的美术老师评为最差，并因此使我一年的绘画科目不及格。两年后，这张画在海淀区的中学生画展上却为我赢得了第一个绘画奖项。环境的改变会使观察风景的视角，审美观乃至评价体系都会发生巨大的变化。如今我依然向往着儿时那种自由的涂鸦状态，那时更关注于自我的心境，有热情和想象力，至于画画的形式、风格以及周围人的评价，则很少在意。其实只有在这种状态下画画抑或做设计，才可能产生乐趣：沉浸在自己的梦幻世界中，自娱自乐，这是一种更纯粹，更令人向往的境界，绘画的技巧与表现形式则是次要的。这一点上，我也与父亲的境遇截然不同，他多半时间都在为业主作图，就像是命题作文，时时体现的都是业主方的要求和形式，很少有时间进行自我情感的表现。相比之下，我更希望在一种自由的状态中继续画下去，至少能将这种“画乃吾自画”的状态尽量延长。当然由于不断地迁居。我小时候好多的涂鸦速写作品与我在北科大附中精心制作的航模一起散落在了江苏、北京、陕西、加拿大的“旅途”中，不过有关这些创作的记忆并未因此而褪色。

我最初接触水彩、风景画及风景的历史是在镇江枫叶国际学校的课堂上。无论是在镇江还是在温哥华的枫叶国际学校，作为手工制作和美术训练，以及戏剧艺术教育，其课业均提供了更大的自由发挥空间和更多样化的评价标准。其中有许多题目会令我的父亲瞠目结舌，有的甚至出是他的研究生都未曾涉足的领域。比如规划设计科目中的交通规划，城市社会学研究，以及戏剧艺术课程中的文艺复兴艺术史评论等。

当然，每一年度从欧洲游学回来的研究生学长是检验我课堂所学文艺史知识的另一个途径，每一次归国后的学生研讨会我都积极参与，虽然有很多地方我未曾去过，但从文艺史课堂到各类游学研讨会，画室中了解到的相关知识让我觉得这种认识比走马观花的国外观光团的认识更深切、更牢固。我的旅行生活从很早就开始并形成了习惯，从皖南的乡村、江浙的水乡到陕北的高原，以及北京、江南一些美丽的花园，云南、贵州广袤的梯田，我的很多风景创作题材来源于这些风格迥异的旅行经历。在枫叶中学的生活中，这种旅行习惯与学校所倡导的教学方式出乎意料的一致，于是我非常幸运地参与学校的各类考察项目，看到了更多、更出人意料的风景，并有机会从游学归来的学长那里检验我了解到的有关风景、异域风光的信息与知识是否准确。

我喜欢北美繁茂的森林、蜿蜒的溪流、温煦的阳光，美丽温馨的港湾城市和终年积雪的山峰同处一张画面，中间隔着无尽的秋林红枫。这种景象在每一次进城的渡轮、地铁上会看得非常清晰，其风景与青藏高原的那些雪山完全不同。不仅如此，就在温哥华不远的郊外，有更多更为天然的深山峡谷和巨大的北美松林。几乎每个周末，我都会以生态考察、自然探索、社团活动的方式，去体验风景之美与其中的乐趣。当然，即便在我寄宿的小镇社区里，美景也无处不在，夕阳下的足球场，社区的枫林，绿荫，乃至房东太太的花园，厨房，无一不是绝美的风景，它们共同组成了我对加拿大这个国家的美好印象和独特的风景情怀。这种感觉就像我第一次从飞机上鸟瞰黄土高原时，对风景的感悟及由此铸就的风景的情怀是一样的珍贵。

进入大学后，这条“风景之路”还是会成为我生活的一部分，用文艺史的眼光看待真实生活中“上帝”与人的共同作品，并用真诚的感情去体味，记录一次次的意外发现，无论是震撼还是欣喜的感情。虽然在未来的职业之路上我也不可避免地会像父亲一样，要一次次面对设计名义下的各种命题作文、命题风景。但是在我自己的速写小本和手卷中，这份风景的情怀会一直持续下去。

王若愚

2017 年 9 月，于列治文枫叶 KPU 校区

目录

风景写生集——画树第一 ONE

Landscape Sketching-Painting Tree First

图 1

图1：前景大树的练习，钢笔＋马克笔：大树的肌理、层次和光影的表现几乎是每一位想画好手绘的景观师的必修课。这种专门化的练习需要有一定的耐心。我在一周之内，重复绘制了十几遍大树。从疏密、光影、色彩等方面寻求变化，这只是其中截出来的两幅图。

图 2：夏天的滨水丛林：滨水植物配植中所表现的趣味在于乔、灌、草以及水生植物的合理搭配，丰富多样，并易于形成坡岸植物带的梯度变化。

图 3：植物表现

Cherry blossom
(or. Maple for Autumn)
Greenway
Main Road
Pedestrian.
Arbor
Bush Grass

图3

图 4：各类针、阔叶林混合表现，钢笔＋马克笔：2016 年的整个暑假，在父亲的工作室里做了一系列树木的线描手绘和光影练习。采用水溶性彩铅与油性马克笔混合作图，可以很好地保留墨水的痕迹，并使彩铅的颜色充分融入马克的笔触中，这是快速简单的练习方式。

图 5：钢笔线描练习 2

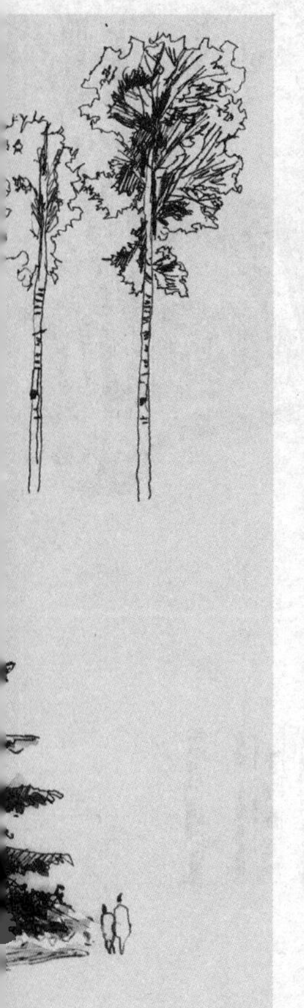

图5

a comprehensive expression of line,
shadow, and layers. 2017.6.
the winter forest,
bark of white birch
图 6

图6：对一棵前景大树的光影、层次、肌理的综合表现

图7：西伯利亚落叶松是最有地域性特征的寒温带植物，这种植物的绘画方式表现得最精细的画家是俄罗斯的希什金。他在画每一张风景油画之初，几乎都会有一张同一场景的素描、速写，都是植物表达的最佳范本。而用美国生产的AD油性马克笔表达这一题材，较之炭笔素描，其自由涂抹的方式可以使作品显现出更灵动的气韵。

The typical Larch (Pine) of Cold Belt,
is the most distinctive Regional Tree.
After the sketch of Russian Painter
Ivan Shishkin.
Ruoyu. 2016.

图 7

图 8

The Alpine Fir forest widely distributed in the Qinghai-Tibet Plateau in China.

图 9

图8：白雪覆盖的山坡，雪景是练习画灰色调的最佳题材，光影是唯一活跃的因素，可以充分利用它们来表现和丰富这类场景。

图9：崎岖山崖上的北美红松

图10：中国青藏高原是高山冷杉林分布最广的地区。画面中巨大的冷杉树林作为前景，映衬出色彩丰富的高山草甸和远处的雪山，营造出近、中、远三个层次的景观效果。

In snow covered slope, light and shadow become the only
active element in the space. Rouyu.2016.

图10

The Autumn Pine Forest
Ruoyu . 2016.
Natural Forest and slope · Ruoyu

图11

图 11：秋叶与柏树轴线鸟瞰

图 12、图 13：秋林植物与山形光影的表现，用线条表现自然山坡的光影与层次。

图 12

图 13

图 14

The sycamore trees in front of Luxembourg palace, paris.
In the light of winter sunset, the shadow so dim and smooth, only the
twisted branch become the main character in the space.

In the Sunny Tuscany Field, everything so shining with golden and brown.
the horizontal skyline contrast strongly with the vertical cypress.
Ruoyu.2017.

图 15

图 14：意大利柏树：记于罗马平乔山瞭望台

图 15：巴黎卢森堡宫殿前的巨大梧桐在冬日夕阳下，光影暗淡而柔和，虬曲婉转的枝条成为整个空间的趣味中心。

在托斯卡纳灿烂阳光下，所有的东西都披上了金色和棕色的外衣，柏树是画面唯一的纵向构图要素，这与水平延展的地面形成强烈对比，相映成趣。

图 16

图 16、图 17：红土城外的薰衣草园，法国普罗旺斯花田。

图 17

图 18

图 18、图 19：在中国古代绘画与文学作品中，松与柏具有相同的象征性含义，而且多以栽松概括。论其本质，情性则长久（寿），坚韧，具有独立人格等。除陵园建设中特别指出种植柏树之外，中国古文献极少出现植柏的内容，都用种松代之。但现今游览北京各大古园时，会发现园中种植了大量有数百年以上历史的古柏，可知，中国古代造园，松柏混种，都以松相称。

Pine and cypress, always hold the Same meaning
in Chinese ancient scroll, they all symbolize the longevity
and other noble qualities as indignity and resilience,

图19

图 20

图 20、图 21: 中国园林中的柏树

柏树是练习线面结合表达的较好题材。通常这类观景的大柏树被称为龙柏。比如北京的中山公园，故宫太庙，都有数百年的古柏。中山公园周边是北京古柏保留最多的地方，此地兴起于辽金时代，当时是辽南京郊的兴国寺。这些古树有的迄今超过八百年的历史，其形颇壮观，虬枝盘曲，生生不息，俗称“中山辽柏”。

图 21

The Ancient cypress in Zhongshan Park (Royal Western Garden in Qing Dynasty. Line and stroke practice by Wang Ruoyu. 2011)

图 22

图 22：中山公园古柏线条与笔触练习

本页三幅龙柏，枝干表达极其简单，主要依靠密集的细线、断线和马克笔相互晕染，形成柏树特有的空间层次。

图 23

图 24

图 23：杭州灵隐寺前的树林

图 24：树干的钢笔素描练习

图 25：罗马蒂沃利的埃斯特庄园里那些长势奇绝壮美的柏树：这座庄园有巴洛克最壮观的水轴和百泉路，这里有高大的植物，错落有致的花坛和各式各样的喷泉，是罗马的微缩景观。不同于欧洲的规则式园林植物，这里的古柏完全自由生长。事实上从 18—19 世纪，除了少数文物学家、考古专家，如德国人温克尔曼到访过，这座文艺复兴时期的庄园一直无人问津，无人打理，尤其是这里的古柏，完全处于自由生长的状态，虬曲环绕，以致使人误以为埃斯特庄园是一座自然式花园。其原始的规则布局样式，可以从法国人杜贝拉克的版画中欣赏到。

The villa garden of d'Este in Tivoli is famous for its
fantastic, natural style of Cypress. Apart from the Avenue
of Hundred Fountains, water Organ by Lorenzo Bernini, and
Mini-Rome which embodied the Ambition of the great Cardinal
Ippolito d'Este, and another hydraulic Wonders, the free style
Cypress is the most incredible work of the Nature, the God.
In the whole 18th, and 19th Century, the villa was abandoned
for two centuries, undisturbed by common clipstyle, free to
grow in its own way, and directly brought a High-Renaissan-
ce style garden in to an atmosphere of Natural forest.
While the origional style can only be found in the Engraving
of French designer Etienne du Perac, which gave the most
precise depiction for the origional Renaissance garden.

图25

图 26：山林的远景鸟瞰

图 27：层次练习 —— 树林的前、中、近景对比，针、阔叶对比，乔、灌、草对比。

图27

"六朝松". The Southern Ancient Pine in the
Campus of DongNan University.

图 28

图 28：位于南京东南大学校园内的古柏“六朝松”

图 29：自然的山坡和林带，将杂树林或松林作为画面的一个重要背景层次，在表现正式的景观效果时会经常用到。

图 29

乡村与田野之美

The Beauty Of The Country And The Field

TWO

图30

图31

图 30、图 31、图 32：诗意景观的风景衍化过程的图像化解说

步骤 1：原始荒凉的海边山崖谷地，光秃秃的山岭点缀着少许的荆棘和先锋灌木，大树无处容身。

步骤 2：在山谷的迎风坡与背风坡开始出现两类显著变化。两种完全不同的生态和风景演化过程开始同步出现。在迎风面，高盐碱的海风将残存的植物一扫而空，以至于浅海滩变成更加裸露的沙丘山岭， 所剩的只有少数的灌木，碱蓬。背风面的变化与此正好相反，这里有丰沛的雨水，在汇水坡一侧率先出现了小树林，能够抵挡强风，并有一定耐盐碱能力的海岸松，在最初的竞争中率先取得优势。在无人干预的自然区域沿山脊排水，谷地缓坡（而非谷底）区域则是林木最丰茂的区域。

步骤 3：人类活动影响下的景观。人类的农业耕作极大加速了风景演化的历史，托斯卡纳是其中一个典范，丰厚的农业传统，多样化的农业园艺赋予大地高度艺术化的景观形态。以位于小山顶部的别墅为中心，周边围绕着多样的葡萄园和橄榄园，外侧则是色彩丰富的各种田地和连绵起伏的草坡（草坡种植在欧洲史上历来被作为一项高收益，高价值产业），形成包含居住与生产的完整的多样化景观综合体。在这一景观综合体中，每一种元素，花园与田园，建筑与风景都高度融合，和谐相处，共同形成我们对托斯卡纳农业景观的印象。

图 32

A Olive tree in Mid-
Italian Country.
The burning Olive trees under brush of
Vincent Van Gogh.
The region is not only rich in high qualit
the red wine of Bruncello di Montalcino or
Montepulciano, the splendid view of Val
the best genuine Olive oil which can be d
period. The gentlely Rolling green hills
and lines of Cypress alongside the count
across the fields, all combine the
Great Val d'Orcia valley
Lines of Olive tree plantation
in Tuscany Augu. 2016.
The Olive tree in Villa
When or who planted the
the atmosphere of this old
back to Ancient Roman
Renaissance Artists,
are

图33

The Olive Trees probably first planted in Asia, Middle-East especially, and then, it spreaded to Mediterranean areas, in ancient Greek, Rome, etc. From then on, it become a very important plant in European lives, a part of landscape in Tuscany or Provence of Southern France. It combines the basic landscape image, with those drought-enduring plants, like Cypress, greaps, in Mid-Italia, especially in Tuscany.

In Val d'Orcia Valley, those famous ~~ancient~~ Renaissance villas, gardens were always surrounded by vineyards, Orchards and olive groves.

The pattern is of beautiful and also practical.

Dream Countryside in Tuscany

The Layout of a Renaissance Garden in Tuscany, "So-called" Ripple Pattern". means, with villa and gardens as the center, and layout outwards with orchards, vineyards and Olive grove, small forest always insert in this pattern, while the outmost layer is the farmland. they all combined the fantastic image of Tuscany Countryside.

Ruoyu. 2016 in Beiji

No one knows exactly
but it truely perfect matches
some of the trees could be dated
ing to clear, It inspired so many
them, Pirro Ligorio, Bramante and Raffaello (Raphael)
Ruoyu in Beijing.

图 33：托斯卡纳地区不仅以盛产高品质的红酒而著称，如蒙塔奇诺的布内罗葡萄酒，以及蒙特布扎诺的维诺比亚葡萄酒等，这片风景绝伦的山谷还以生产最好的橄榄油而闻名，这种壮观的种植园景观可以一直追溯到文艺复兴时期。孤植的橄榄树点缀在棕色、黄色和绿色山峦上，柏树沿着乡村道路的两侧种植，草垛布满整片田野，所有的一切构成了托斯卡纳南部欧西亚的最壮观、最令人称道的梦幻田野景象。

橄榄树，最先在亚洲，特别是中东地区种植，然后逐步扩展到地中海地区，如古希腊和古罗马，日渐成为欧洲人生活中重要的作物，同时也是法国南部和意大利中部重要的景观。在欧西亚山谷，酒庄，果园和橄榄园环绕在文艺复兴时期的村舍周围。它们和意大利中部特别是托斯卡纳地区的耐旱植物，比如柏树、小麦、葡萄等一起构成了托斯卡纳景观的基底图像。

一座典型的托斯卡纳文艺复兴时期别墅的布局，即所谓的“涟漪”布局：将别墅与花园布置在场地中央，将葡萄园和橄榄园围绕在中央别墅四周，其中插入一些小规模的林苑，大量的田野则位于布局的最外侧。所有的元素结合构成了迷人的托斯卡纳梦幻乡村风景画。

位于罗马蒂沃利的哈德良别墅的橄榄林，没人能确切知道这些古老的橄榄林由谁种植？何时种植？但它无疑与这座巨大的罗马别墅构成了完美和谐的画面。其中很大一部分古橄榄林可以追溯到古罗马时代。可以确认的是，这些橄榄林与这座哈德良别墅的花园建筑一起激发了无数文艺复兴时期的艺术家、建筑家的灵感。其中的皮埃尔・佛朗西斯科・菲奥雷迪诺、利戈里奥・皮罗、多纳托・伯拉孟特和拉斐尔・桑齐奥是佼佼者。

图 34：梯田的色彩。云南的梯田在最美的三四月份会呈现出极为绚烂的色彩光影。此时的山间有溪流，有植物生长，水田中会出现明暗不同的色彩变化。在去安龙参加第二届安龙可持续发展与设计国际夏令营之前的几天，除了查询必要的设计资料，其余的时间几乎都用来画类似的光影手绘稿，这是其中的一幅。

图 35：涉及光影与色彩的题材，首先想到的是文森特・威廉・梵・高在普罗旺斯的阿尔小城留下的故事。那些充满激情的笔触与渲染手法，极有借鉴意义的农田景观绘画，以及天空的灰暗与金色田野之间的对比，加深了这种金色麦田的表达效果。

图 34

图 35

Dream land of Tuscany
Val d'Orcia Valley from
the city wall of Pienza
2016. in Beijing.

图 36

图 36：从皮恩扎古城的城墙上看到的欧西亚山谷

图 37：在贵州安龙举办的第二届安龙可持续发展与设计国际夏令营中，我试图再现这种农业种植之美，便试画了几张梯田的手稿，钢笔与毛笔混用。

图 37

Exploring the rolling wheat covered (or, sunflower fields) of Southern Tuscany. When you up to the hill following
the cypress line across the ancient Roman Road, you see the gentlely rolling green, yellow, brown and those smoothly
streached shadow spread across all the small hills and fields, with some lonely cypresses at the top, and rolls of
haystacks spread across the fields. Beside these are the breathtaking landscapes which demonstrated the Valley of Orcia
as the ancient site dated back to Roman time, medieval castles, hilltop towns and charming farmhouses, rows of
vineyards and patchs of olive gardens, which all manifested the basic rule alongside the fantastic view, that is,
Beauty is always practical..

图 38

图 39

图 38：托斯卡纳连绵的麦田与花田

沿着柏树指引的罗马古道登上小山顶，可以看到交相辉映的，温柔起伏的托斯卡纳田野，那是欧洲风景中最值得夸耀的梦幻田野与乡村。从皮恩扎古城墙上能看到典型的欧西亚山谷景观，绿色，棕色的田野和山坡，拖着长长的如丝绸般柔滑的光影，一卷卷的稻草垛散在山前坡脚，如今这些就是守着通往罗马大道上的驿站。中世纪的城堡，山顶的石头小镇，迷人的农场风光，连绵的橄榄园和葡萄园，这些美得令人眩晕的风景，似乎都在向我们暗示：风景之美一直与当地的经济，生活相伴而行。

图 39：从大教堂一侧的山坡上看锡耶纳城市天际线。

图 40：蒙地切罗城外的乡村风光

托斯卡纳的圣奎里科与蒙地切罗是托斯卡纳最著名的葡萄酒产地，也是中世纪以来，基安蒂酒业最出色的代表。

San Quirico & Monticchiello 奎里柯 & 蒙地切罗
(A small medieval walled village to south of Pienza, presents the great view of the dream land of cypress lined roads.)
A olive tree in Mid Italian Country.
Rural Cypress Road near the Tuscany Town of Monticchiello.

图40

图 41

图 41：欧西亚山谷中的乡村田野和风景小路。托斯卡纳的乡村风光多由这种蜿蜒的小路一一展开，这或许是所谓的“风景绿道”最具启发性的源头，这条梦幻的乡村之路，一路向北延伸，通往著名的中世纪小城，蒙塔奇诺，蒙特布诺，皮斯托，圣吉米亚诺，一直将你带回托斯卡纳省的首府佛罗伦萨。

通往美丽天堂的梦幻之路：托斯卡纳乡村小路风景中，最突出的元素是那条或笔直，或蜿蜒的巨大柏树轴线，与那些独自挺立的巨大山顶柏树不同，这种由千百棵意大利柏组成的轴线，几乎可以将人的视线引向极远的天际，直达一座中世纪的城寨，古堡。

这些强大古拙的柏树似乎在暗示着后来绵延于全欧洲数个世纪的巴洛克轴线景观的真正来源。

图 42：山岭上孤独、苍翠的柏树是连绵风景的连结点与唯一的垂直景观。

山顶上的居住单元之一——别墅，道路边上的柏树轴线，别墅周边的橄榄，种植园和远方连绵的远山共同构成了梦幻乡村的完美天际线。

图 43

图 43：托斯卡纳的庄园农业景观。远方山岭上的庄园，由红顶、白榉和深绿色近乎黑色的柏树、果林构成，其周边围绕着葱郁的葡萄园，橄榄林；外围则是一望无际的美丽田野，最终形成由“庄园——花园——果园——葡萄与橄榄园——外围田野”组成的层层发散的同心圆结构的独特景观。

图 44

图 45

图 44：普罗旺斯 - 阿尔地区的乡村风光——麦田边的小农庄。

图 45：麦田系列——金色麦田中的鸭群，是米勒，库尔贝和很多枫丹白露派的画家热衷于表现的风景。

图 46：斑斓的田野，山谷和远方的蒙特布扎诺古城。

A distant view of Montalcino in
Tuscany
the Medieval town of
Montepulciano in the distance

图 46

图 47

图 47：文森特 · 威廉 · 梵 · 高笔下燃烧的柏树，扭曲、跃动的笔触，金光闪闪的田野，村庄，劳作的男女共同构成普罗旺斯－阿尔地区的乡村风景。

图 48：基安蒂酒庄的典型风景——连绵起伏的葡萄园。

图 48

图 49：三座山脊上的中世纪城

锡耶纳坐落于三座山脊之上，三座小山上所有的建筑延狭窄的街巷展开，而街巷的终点无一例外则是市中心的田园广场。城市的各种景观由这一点伸展出去，并最终由多条道路重新被聚拢到广场中间。

托斯卡纳的小城锡耶纳的天际线大体由灰白色的主教堂和砖红色的市中心田园广场组成。大教堂的穹顶和钟塔以及田园广场的曼吉亚塔楼，是天际线上最突出的三个标志物。相应的，我在小城的速写和旅行也围绕着这几个标志点展开。这座中世纪的古城实际上位于三个相互联接的山脊之上，探索其中的空间需要不断地上山，下坡，再上山：所有的行程转换都会回到城市的中央广场——田园广场，空间行进中所有的画面联接成对这座中世纪山城的全部印象。空间的叙事性和多样统一性在这座城市中形象地反映出来。

Three parts of the Medieval City,
Siena, perched on three ridges, while the
intersection of the three is the center of Siena,
Piazza del Campo.

图 49

The skyline of Siena combined with two parts,
Duomo of the town, the red Palazzo Pubblico and
its tower, Torre del Mangia, (曼吉亚塔楼) Accordingly,
the search for the city's main attractions also base
on these three ridges, always goes up and down to
connect all these sceneries, and makes full image of
the medieval town.

关于旅行速写

旅行是一种让人脱胎换骨的体验，旅行中的一切都是全新的体验。新的环境中，视野在扩大，视角在不断改变。几乎每一次旅行都能教会我新的东西，使得平常很容易被忽视的东西变得丰富有趣。平日里若要创作一幅画，也许会花很长时间构思，修改，最终往往搞得激情全无。而在全新的环境和独特的体验中，创作的机遇往往会不期而至，甚至俯拾皆是。如果说相机所记录的旅行更像是事无巨细的客观世界，速写所展现的旅行则更多的是独特视角下的个人感受与情趣。

相对于走马观花的拍照式旅行，速写本质上是观察所得，画面的构成极大地取决于个人的兴趣与取舍。这就像中国古人所说的“中得心源”，是从激起个人情感的万千细节中，发现一个崭新的世界。相较于瞬间的摄影，这种记录方式更能构建永恒的记忆。其过程往往是发现兴趣点，停下脚步，慢慢品味，然后用眼和心努力去感受，最后带着激情将这些感受记录下来（其表达形式甚至不在乎是图形还是文字）。这种情况下，那些与表达主旨并无直接联系的细节被甩在一边，所见，所记录下的皆是心灵之洞见，与真实场景的客观性之间的关系其实不那么密切，你所记录的是你眼中的世界。你可以从自己寥寥几笔的速写中了解当时所处的环境，甚至通过这些草图回忆起当时作画的心境，以及人物、场景是否嘈杂等，所以即便是那些在别人眼中无法理解的乱涂乱画，在作者眼中却能幻化出一连串的回忆与联想。同样，你却很难从一堆旅行照片中回忆起某一次旅行的种种细节，尽管这些照片看上去更加细致全面，但正是这样的事无巨细，反倒是缺失了作者本人的体验，进而使一次旅行变成一种毫无特色的走马观花，这其实是大多数旅游者都有过的真实体验。所以要真正“看见”那些旅行中的风景，需要的是手中的笔，而未必是一台相机。这就是为什么针对同一景点，每一张照片几乎都大同小异，而同一景点的速写却可以因人，因心情而显得各不相同；对同一场景，每一个人通常都会因兴趣差异，而发现完全不同的独特之处，并加以人为的刻画渲染，使这种特色进一步强化。对旅行所见，诸如城市的人流，天际线，乡村屋舍，田野以及荒野的岩石，杂草，锈迹斑斑的门窗，由绘画表现出的情味是各不相同的。这才是真正的、属于个人的旅行记忆，它远远抛开了千篇一律的旅行手册和喋喋不休的导游词和事无巨细、漫无目的，如机枪扫射一般拍摄的海量照片，以极少的笔墨和图像表现更为丰富的记忆与场景感。

旅行中的体验因记录方式有差异而显得不同。这种不同从旅行的第一天就会显现出来。比如一大早起来，挥舞着导游册，匆匆赶出去迎接日出的往往是那些举着相机不停拍摄的观光客，而准备记录这段旅行的人，却总是坚持要把他们的第一顿早餐画下来，然后欣然走出旅馆。早餐桌上的风景是每一次旅行深入观察并理解新世界的起点：桌布，餐具、咖啡具，焦黄的面包，熏肉，以及一旁厚实的壁炉架和墙上的装饰等，这些精美的细节极大地影响着你对所要描绘的新环境的整体印象，而这些又是那些匆匆忙忙，左顾右盼的观光客难以察觉的。环境的巨大差异，新鲜感的驱动是一种无与伦比的引导和体验，足以调动你所有的感官，这是感悟风景，记录旅行的第一步。

说来奇怪，我的第一次旅行并不是如人们通常计划的那样，去选择一处风景优美的地方或是一个历史文化底蕴深厚的城市，而是在一连串大雨的夏日，跟团去了香港、澳门。那一次还没有速写本的陪伴，所有的记忆似乎是在不断上车，下车，躲雨，拍照，和熙熙攘攘的街市上度过的。同行的大人们都在忙着一家家地购物，而我只是傻傻的站在一边。这次旅行对我而言，除了香港、澳门，这几个空间名称，几乎没有留下任何属于自己的记忆。

最开始用速写记录旅行是在初中时，一次周末的郊游，我围绕北京周边的园林进行了一些简单的速写。这或许称不上真正的旅行，但却是我速写风景的最早实践。颐和园后山的白皮松，中山公园的古柏，西山八大处的秋叶是我经常选择的速写题材。每次为了画好一棵树而在父亲设计室的画桌上苦苦临摹的日子是枯燥的，而把这种锻炼放在夕阳下，大红宫墙外的中山公园或幽静的颐和园后山，日子就会轻松而有激情。这些画树的经历如此美妙（虽然当时的作品未必件件都尽如人意，有的甚至差强人意，但这又何妨？）在后来去皖南画石板路，去山西画大院，以及去西藏林芝等地去描绘那巨型的冷杉林的时候，这些乱涂乱画的基本动作都轻松地派上了用场。西藏的旅行进行了半月之久，而真正感动我的并不是那些想象中的拉萨的寺庙和那些磕着长头的藏区各地的虔诚的乡民，或许是我对这些场景期待已久，真的到了眼前，这一切反而变得习以为常了。而那些完全意料之外的，甚至是无法回避的艰难和奇遇，反而能激起我对此次旅行更为强烈的欲望。比如顶着不期而遇的暴雪在大山之中驱车，亲身体验那些令人终生难忘的场景：怀着巨大能量的雪松、冷杉、牦牛，那些具有超乎想象的韧性的雪域生灵，在风雪之中展现的生命张力。我曾描绘过苏格兰的红松，加拿大的雪松，但西藏的冷杉与这些巨木截然不同，他们几乎全部生长在雪线之上，冬天里，这些参天巨木几乎全部被大雪覆盖，其分枝与叶片全部减少，而将能量全部集中于主干，似乎要从地层的深处直冲云霄。这种冰天雪地里的顽强生命给我极大的震撼。我在近乎冻结的越野车上，哆嗦着画下了它的身躯轮廓，并在赶回八一镇的当晚，就着旅馆的青稞、藏粑和依旧鲜明的记忆，完成了对这些巨木林海的描绘。这些是真正能令我回想西藏，回想起雪域的景象。

徽州是一处令人难忘的江南风雅之地，即使和着乡野的牛羊，鸡棚，也并不妨碍你感受这份儒教之乡的文化气息。从黄山下来，经深渡、宏村一路走到新安江畔，第一次感受到边走边画的激情。感受的前提是避开喧闹的旅游人群，尽可能多地选择背街陋巷。这里的石板依旧留着苔痕，阳光从狭窄的一线天中洒下，落到石板上，已变得非常柔和，恰好能映射出小巷的斑驳与沧桑，构成了一种非常奇特安静的空间。行走于其间，非常像在日本茶庭的感受：经过一系列的树木障景，汀步（犹如皖南小巷的石板路），期待转折处的美丽风景——利休大师的茶室。在这里，或许是“大夫第”的永恒的砖雕门楼，或许是转过大门后厅堂梁架上那些精美的月梁和雀替。这种期待从踏入小巷石板路的那一刻开始，经过一道道砖雕门楼、老屋、火弄坑，一步一步接近终点，路上每一个老物件，都像是为了构成这部起承转合的序曲必不可少的铺垫。

或许徽州的古老都凝聚在了一条巷子里，这条巷子，和我家乡的巷子完全不同，当然这不是我在温哥华的社区小路上所能感受到的。但无论多远，每次只要展开徽州的那卷图轴，徽州街巷里所有的印象，甚至是当时的点滴细节，都能源源不断地流到我眼前，这或许就是旅行速写真正令人感动之处，也是它真正的价值所在吧。

王若愚

2017.10 于温哥华 KPU 枫叶校区

旅行速写 Travel Sketch Book THREE

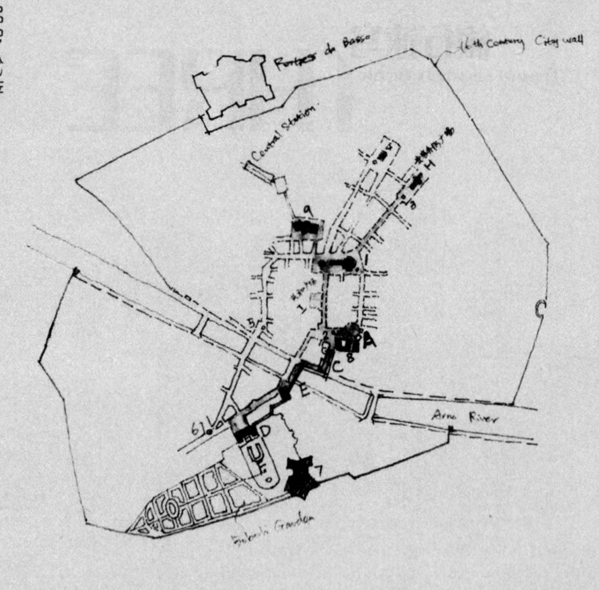

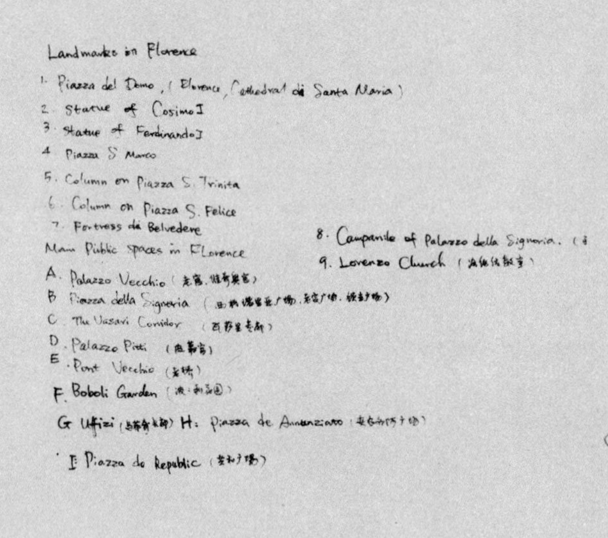

图 50

图 50：西格诺利亚广场总体空间分析

A 兰奇敞廊；B 维奇奥宫入口；C 维奇奥宫（老宫）；D 乌菲奇长廊；E 海神（尼普顿）雕塑（巴托洛米奥·阿曼纳蒂的作品）；F 大公科西莫一世骑马像（朮韦努托·切利尼作品）；G 米开朗基罗·博那罗蒂的大卫雕塑复制品 H 朮弟里尼雕塑

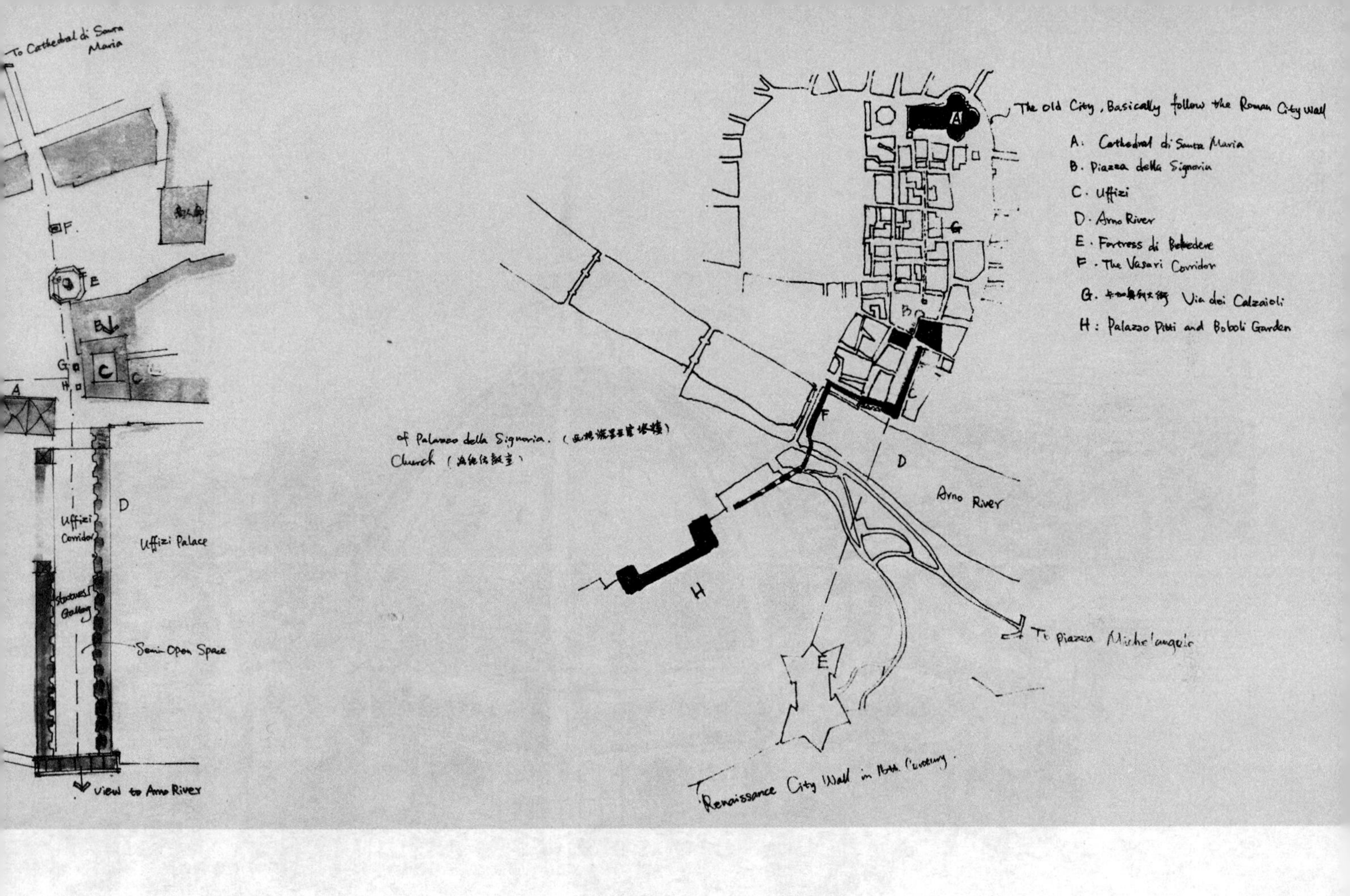
To Cathedral di Santa Maria
F.
E
B
G
H
C
C
A
D
Uffizi Corridor
Uffizi Palace
Statues Gallery
Semi-Open Space
view to Arno River
The old City, Basically follow the Roman City Wall
A: Cathedral di Santa Maria
B. Piazza della Signoria
C. Uffizi
D. Arno River
E. Fortress di Belvedere
F. The Vasari Corridor
G. Via dei Calzaioli
H: Palazzo Pitti and Boboli Garden
of Palazzo della Signoria.
Church
D
Arno River
H
E
To piazza Michelangelo
Renaissance City Wall in 16th Century.

图 51

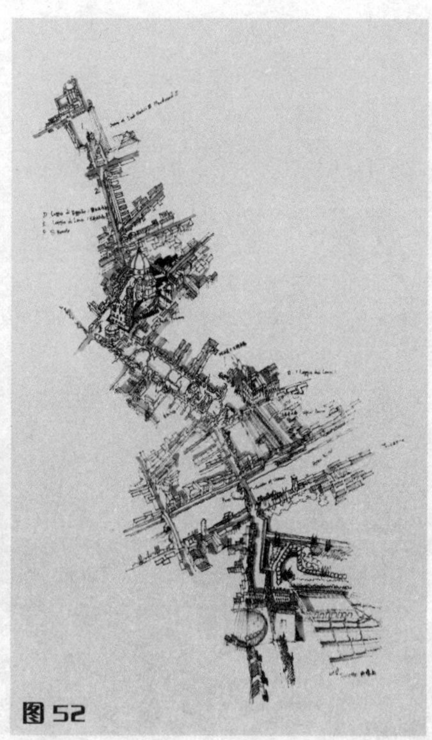

图 52

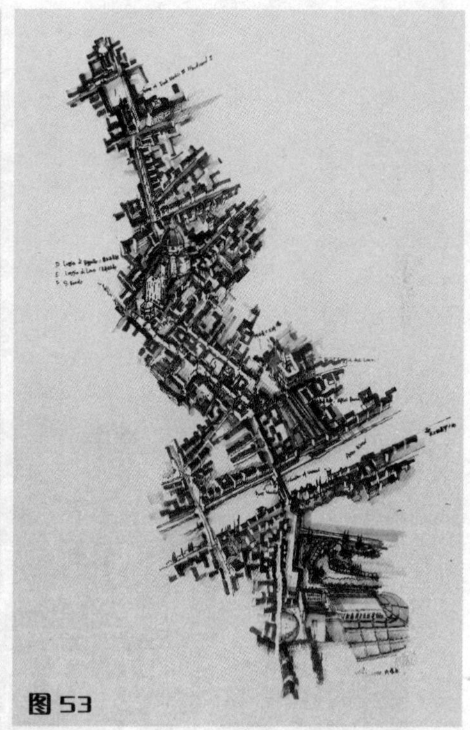

图 53

图 51：佛罗伦萨大教堂穹顶鸟瞰（从乔托钟塔鸟瞰全城）

图 52、图 53：佛罗伦萨历史街区鸟瞰图

A Bird-View of the City Florence from Boboli Garden
Wang Ruoyu. 2017.7. Beijing.

图 54

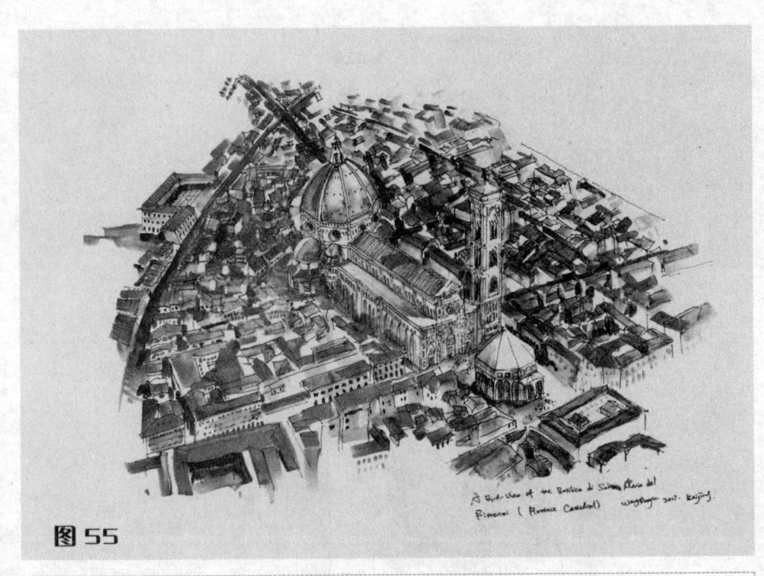

A Bird-view of the Basilica di Santa Maria del
Fiorenze (Florence Cathedral)
Beijing.

图 55

图 56

图 54：从波波利花园鸟瞰佛罗伦萨圣母百花大教堂

图 55、图 56：佛罗伦萨大教堂鸟瞰

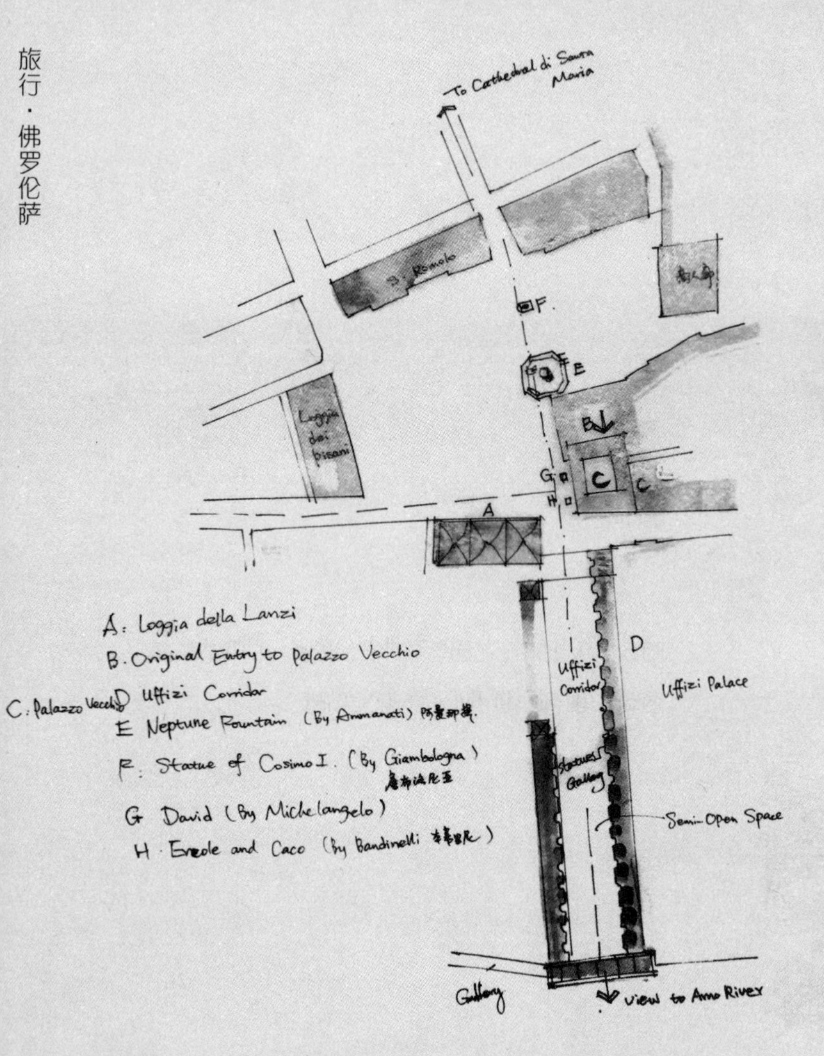

Analysis of Piazza di Signoria.

图 57

inside is the Lanzi
the treasure of Rena
and everyday, her
Crowded place
People come he
of Cellini, D
to see the
Florence Statue of
Perseus with the head of Medusa at Piazza di Signoria.
By the great master Cellini in Early Renaissance.
Actually, the great defeating of the exotic Monster by
Perseus, exploits the manifesto, a testimony of Great
Similarly, the marble statue of David by M
standing on the front of Cellini's
Same meaning, one of
Stories

Michelangelo has been preserved in
Academy Museum. Ruoyu.
by Lorenzo the Magnificent in
of 15th Century. Ruoyu. 2016.

图 57：兰奇敞廊

经历了 8 个世纪风雨的兰奇敞廊是佛罗伦萨老宫广场上最具特色的公共艺术空间，是一座不折不扣的文艺复兴雕塑艺术宝库，也是如今佛罗伦萨最拥挤，受关注最多的旅游景点。那些文艺复兴早期和盛期的大师作品：切里尼的帕修斯像，大公科西莫一世骑马像，阿曼纳蒂的海神喷泉，还有最为重要的作品，那座作为佛罗伦萨标志的大卫雕塑。尽管这是复制品，真正的原作现存于佛罗伦萨学院美术馆，但当所有的杰作并置在一起时，其感受往往是出人意料的。佛罗伦萨西格诺里广场最著名的雕塑，帕修斯高举梅杜莎的头，由文艺复兴早期的金工大师切利尼所做。事实上，通过对斩杀异域妖魔的希腊英雄（帕修斯）的描绘与颂扬，该场景成为一种隐喻，甚至是宣言，标志了伟大的美第奇家族的胜利。同样，广场一侧的巷官门前，耸立着米开朗琪罗的杰作大卫（复印品），与此具有相同的隐喻，这些创作于 15 世纪后期的伟大作品，是美第奇家族一系列神话隐喻的重要作品，其委托人便是家族最伟大的赞助人洛伦佐 · 德 · 美第奇。

图58

图 58：从小巷中看维奇奥宫及广场空间

图 59：罗马埃斯特庄园里有自由生长的迷人的柏树，埃斯特庄园除了拥有举世闻名的百泉路，水风琴（贝尼尼作品）和“小罗马”模型建筑群以外，园中柏树的轴线，树林及著名的柏树剧场（cypress arena）等景观堪称是最令人难以置信的大自然造物。

在 18—19 世纪，这座别墅从辉煌的顶峰跌落，逐步走向荒废，以至于两个世纪中，几乎无人问津。园中柏树在自然界中不受任何束缚，自由自在地“疯长”，形成一种极为壮观，极具原始韵味的“自然林区”，而中世纪的规则式布局种植园，几乎被完全掩盖，以至荡然无存。如今，埃斯特庄园的原始布局样式只能从法国园林大师埃蒂安·杜贝拉克的版画中得以一见。

图60

图61

图 60：巨大雄伟的柏树轴线是文艺复兴盛期埃斯特园林最杰出的景观特色。

图 61：埃斯特庄园的柏树（炭笔灰卡纸练习稿）

图 62

图 62：埃斯特庄园的柏树：在经历了 16 世纪后半叶最辉煌的时代后，这座位于蒂沃利的埃斯特家族最伟大的园林走入了一段近两个世纪无人问津的荒废时代。这期间，除了极少数来自异国的古物专家（或称文物贩子），如德国人温克尔曼等人，偶尔会光顾，这座花园在无人干扰的情况下，自由自在地生长。这些古文物专家从庄园中移走了大量罗马时代的雕塑和古物，却对庄园中姿态奇绝，千载难逢的大柏树不屑一顾。也正因为如此，自然留给这座庄园最珍贵的遗产得以保留。

在所有的有关于埃斯特庄园的名人中，最著名的音乐家李斯特在园中谱写了他的名作《埃斯特庄园中水的游戏》，并在 1879 年，于园中举行了他一生中最后一次交响乐演奏会。毫无疑问，园中数以百计的喷泉，洞窟、雕塑、水戏，成了音乐家灵感之源。同样，音乐家的作品也是这座 400 多年历史名园最美的总结之作。

Villa d'Este at Tivoli, from the Cypress Avenue,
This is the most critical, and also actually the best
viewpoint to depict the character of the garden. From this
point, the view is gradually raised with step, hedges, Fountains (Dradgon Fountain),
theater (Cypress theater), Statues (once erected here in 16-17th Century), and all
the way down to the top of the villa. The Axis is like a huge channel, and
splited by the Giant's Axe, open the curtain of the great open theater.

图63

图 63：埃斯特庄园的景观轴线

由低地入口处沿着柏树轴线的引导逐层向上，一直延伸到别墅的顶端，观者沿着层层向上的台阶，犹如行走在树篱构造的峡谷之中，视线被直线指引，形成极其壮丽的视觉和舞台感。一字划开的中央轴线两侧丛柏林密布，整体看，犹如是巨人用巨斧直接劈开留下的人类印记。

图 64

图 64：平乔山上罗马散松构成的罗马城杰出的天际线轮廓

The cypress tree of Liao Dynasty in Western Royal Garden. Ruoya

图 65

图 65：帝京千年一望中——北京园林线描

图 66：北海公园的辽代古柏

图 66

The 500-year-old cypress tree in Imperial palace of Forbidden City.
he Ancient Pagoda Tree in Royal West Garden ("Xi Yuan"
or Beihai Park in Modern Beijing) .
图 67

图 67：皇家西苑的唐代槐树（北海公园画舫斋的“唐槐”）/ 故宫 :1000 多年历史的古柏

图 68：北海公园后山的垂花门 / 故宫御花园后苑的垂花门

筒瓦
Hanging Lotus Gate in Royal Western Garden (Modern Beihai Park)
"Chui Hua Gate".

图 68

图 69

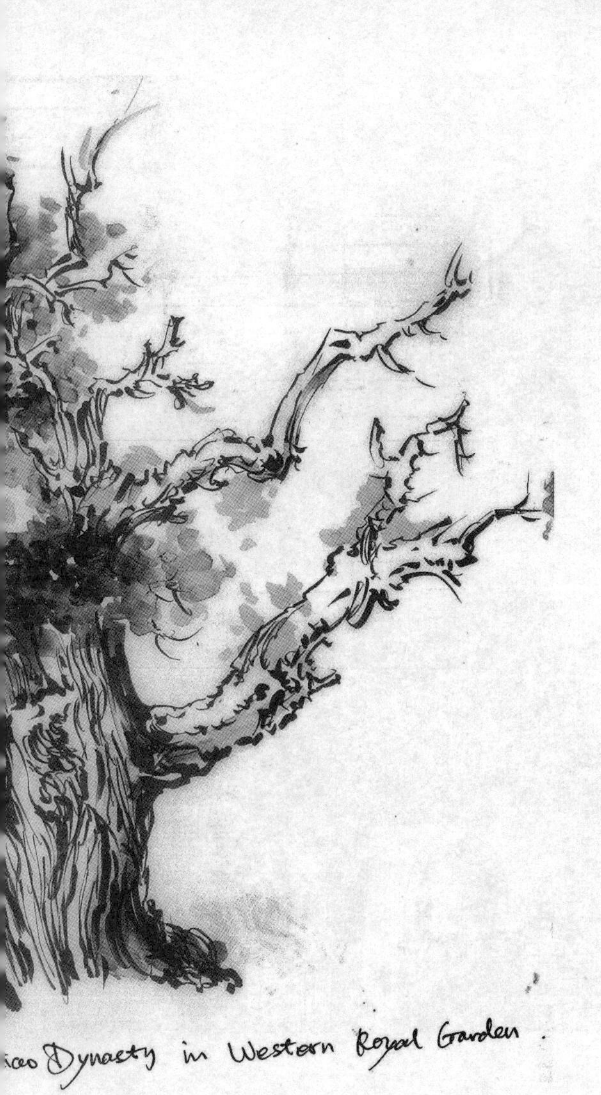
Dynasty in Western Royal Garden.

A bird view of old
Beijing City
2015.

图 70

图 69：故宫天一门：皇家西苑中的古柏。西苑，太庙，社稷坛一带的古柏种植于辽代，是当时皇家北郊御苑骑射休闲之地，后人以“辽柏”称之。

图 70：帝京千年城，关河一望中

White Elephant in the
Imperial garden. WanRuyu.2013.
Beihai Park. ("Xiaoxi Tian)
The Imperial Garden in
Forbidden City
Ruyu
"TianYi" Gate in
Imperial Garden

图 71

图 71：故宫后花园的象与红墙 / 北海公园“小西天” / 故宫天一门入口速写

图 72：宫墙之外——老北京鸟瞰——故宫“西六宫”到御花园一带：故宫“西五所”改建而成的建福宫西花园是乾隆时代内廷花园的典范之作。乾隆登基后，对其进行翻修扩建，成为时常游憩的宫苑园林，此后，这座花园一直得到修缮。此花园一直完好保存直到清代末期，溥仪皇帝在位期间 ，花园毁于一场神秘的大火。

图 72

Wang Ruoyu 2016.5 Tsinghua
The Ancient cypress in the Imperial College ("Guo Zi Jian")

图 73

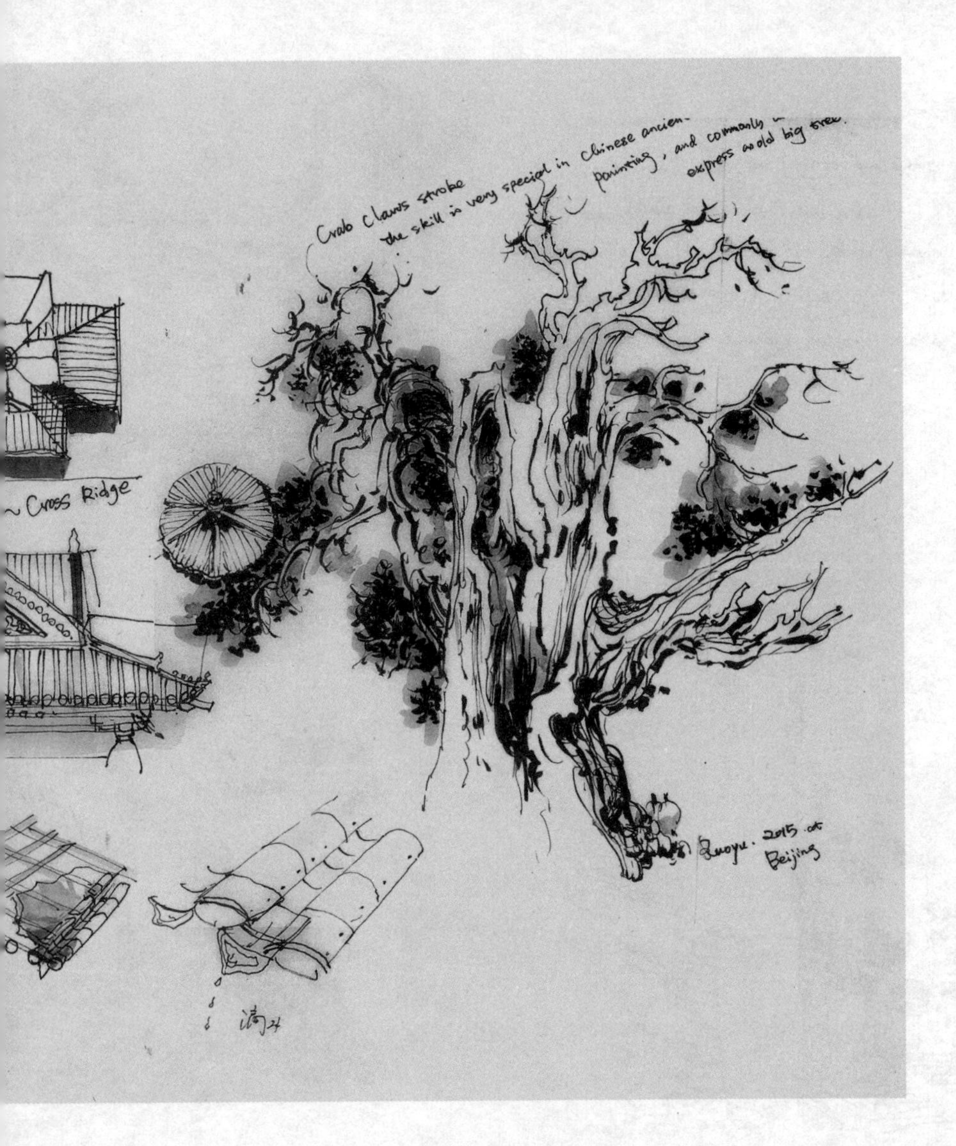

图 73：北京国子监的古柏与古建筑：国画线描中的“蟹爪皴”法，可以用来形容古柏的姿态。

Civilian Huizhou Residential housing embraces typical physical features of Fengshui concepts, Confucian ethic and order, and also, aesthetic ideas. With black tiles, white walls, (so-called horse-head Wall), and other elegant decorations with wood, brick and stone, the housing structure appears full of a sense of beauty. As a school of traditional Chinese architecture, Huizhou architecture amasses elegance, simplicity and dignity in perfect harmony.

图 75

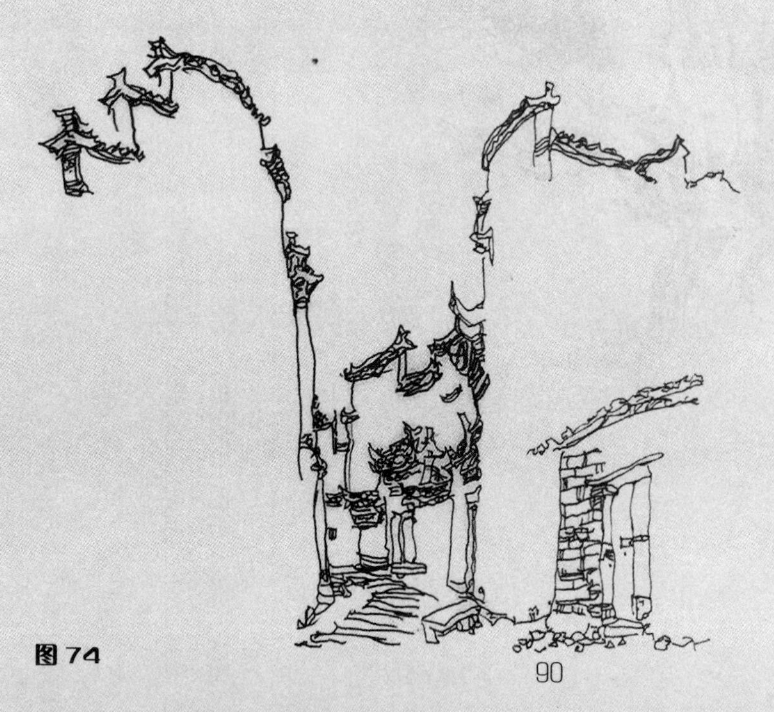

图 74

图 77

图 76

图 78

图 74：徽州小街巷的主要特点是：纯灰色的背景墙面，起伏的马头墙和精致的入口门楼装饰，这些共同构成了一个别致而和谐的居住环境。

图 75、图 76、图 77、图 78：古朴静谧的街巷与精致的入口相映，在形式，理念，自然与人共同建立的空间秩序以及社会伦理等方面体现出一致性与和谐感。

徽州民居

风火墙是徽派建筑的代表，其墙体高于屋面，并层层上升，直到正脊处达到最高，这种高低起伏的山墙最主要的功能是防火防盗，可有效地阻止盗匪从建筑屋顶之间穿梭。

A Horse-head Wall may seperate two parts
of a house or different houses in differen level,
may effectively prevent the fire spread.
The stage and its attached plaza
is the most important space in Huizhou and Zhejiang, The Famous Chinese
writer Luxun has described in his novel this kind of rural stages with the
typical Architecture style of Huizhou.
2016.7.

图 79

图 79：徽州村落中的戏台：徽州以及江浙一带乡村中，最重要的公共空间就是戏台和公共祠堂，而戏台最具富有浓浓的乡情。鲁迅先生在他的作品中详细描述过这种乡村社戏空间的构造特色和使用方式。小巷是徽州民居最具特色的空间 ，每一座徽州民居院落与住宅都由这种狭窄，幽长的小巷相互联结，形成一种高密度的居住结合体——村落。光线从小巷一线天似的上空射入，由上到下逐步减弱，营造一种柔和、温馨的氛围和半室外空间的意象，犹如在各居住单元之间形成的联结与过渡体，也是使用频率最高的公共活动空间。阳光从顶部射下，在石板路上打下长长的投影，并在屋檐，马头墙上勾勒出一道道的银边。

图 80：明清之际，江南盗匪猖獗，对富庶的徽州村落袭扰尤甚。故逐步发展出此类杰出的兼具美感与防御功能的徽州之墙。黑瓦、灰砖、白墙形成的韵律感立面。风火墙亦称“马头墙”，作为徽州民居重要的防火设置，有效防止了木质梁架因着火而蔓延。同时，将不同层次、高度上的屋面一一分割，联结成一个整体，成为徽派民居显著的聚落形式。

Hui zhou Dwelling House. 2016. 9.
e most brilliant element of this residence house is so-call Wu Feng
nclosing Wall, (五凤墙. or 风火墙) or five-phoenix wall, which is much higher
than the houses. The top of the wall goes up step by step. The main Function
of these walls is to ward oft the wind and fire, as well as to prevent the
burglars climb up to top of the house. They also knowing as "wind-and-fire walls"
"Horse head wall"
very small window because of frequent burglars
The wall of Huizhou.
Rhythm of Black, Grey and White
The wall of Huizhou
Ancient Residence

图80

图 81：作为一种生活的空间，小巷装满了徽州人的各种生活记忆。直至当代社会，徽州人一直在这狭窄温馨的小巷中注入自己对生活的美好向往。温柔、朦胧的光从小街上空洒下，勾画出徽州民居的种种细节，也充满了真实生活中美好的情味。

图 81

distinctive
mplex/dwelling
ar lanes forms
arrow
sity
While
ight will
the
lge of eaves
ace, the lane
ries of
in ancient
iety.
M. 2016.
The Gentle, fuzzy light, depicts the details
of the Wall, everything here, is an typical atmosphere,
Calm, elegant and self-confident. A soulful rendition, a
true honest to their life.

大夫第
One of the most Famous Local Residence, (in Hong Cun)
"Da Fu Di", (大夫第) means the House of
high rank official. It is very likely to be the best and
most well-preserved Dwelling House in the whole county
(Xidi, Hong Cun)

图 82

图 82：大夫第砖雕门楼：这是徽州民居中最著名的一个砖雕门楼，“大夫第”意为公卿大夫之宅门，是徽州民居砖雕中保存较好，装饰最精湛的一座，位于徽州西递宏村。

图 83：徽州民居后院花园 / 徽州民居前庭院——“四水归堂”的徽州院落：徽州民居的前院和中庭四檐均向内院倾斜，四檐所收之雨皆不外留，而是被收集于院中央小水池或水缸中，以备日用，称之“四水归堂”。古人视水为财，故“四水归堂”也是取财不外流之意，亦是对生活与未来的美好祈愿。

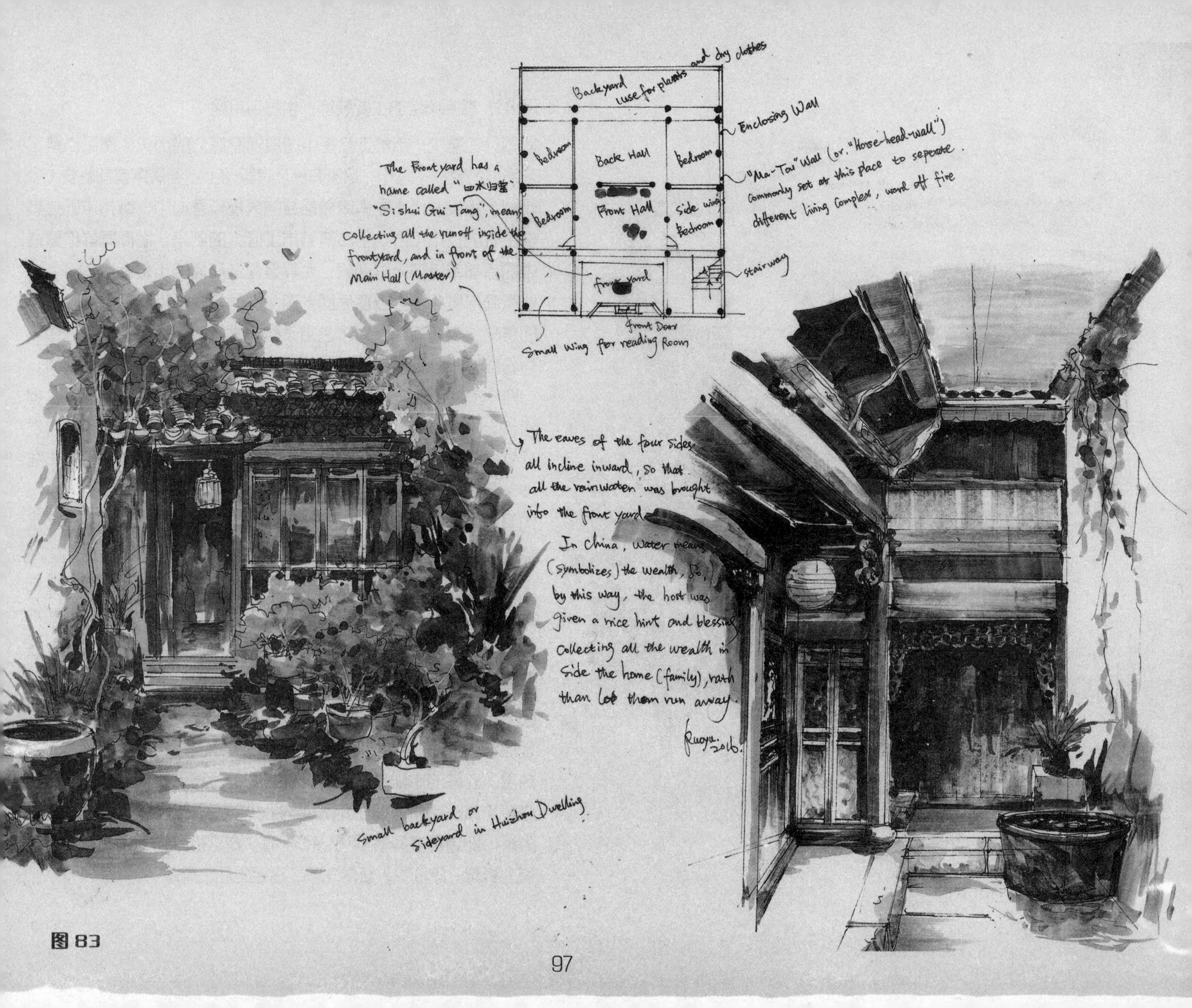
and dry clothes
Backyard
use for plants
Enclosing Wall
Bedroom
Back Hall
Bedroom
"Ma-Tou" Wall (or, "Horse-head-wall")
commonly set at this place to seperate
different living complex, ward off fire.
The Frontyard has a
name called "四水归堂"
"Sishui Gui Tang", means
collecting all the runoff inside the
frontyard, and in front of the
Main Hall (Master)
Bedroom
Front Hall
Side wing
Bedroom
stairway
front yard
front Door
Small wing for reading Room
The eaves of the four sides
all incline inward, so that
all the rainwater was brought
into the front yard.
In China, water means
(symbolizes) the wealth, So,
by this way, the host was
given a nice hint and blessing
collecting all the wealth in
side the home (family), rath
than let them run away.
Ruoyu. 2016.
Small backyard or
Sideyard in Huizhou Dwelling

图 83

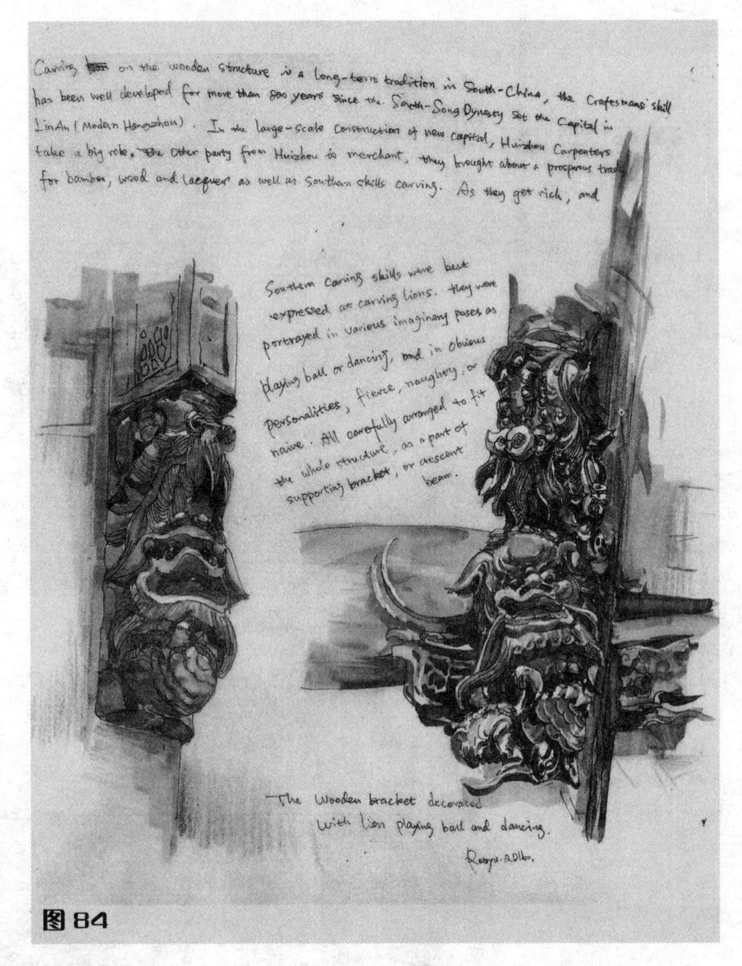

Carving on the wooden structure is a long-term tradition in South-China, the Craftsmans' skill
has been well developed for more than 800 years since the South-Song Dynasty set the Capital in
LinAn (Modern Hangzhou). In the large-scale construction of new capital, Huizhou Carpenters
take a big role, The other party from Huizhou is merchant, they brought about a prosperous trade
for bamboo, wood and lacquer as well as Southern skills carving. As they get rich, and
Southern carving skills were best
expressed at carving lions. they were
portrayed in various imaginary poses as
playing ball or dancing, and in obvious
personalities, fierce, naughty, or
naive. All carefully arranged to fit
the whole structure, as a part of
supporting bracket, or crescent
beam.
The Wooden bracket decorated
With lion playing ball and dancing.

图 84

图 84：徽州木雕的主要装饰：牛腿与月梁

徽州“三雕”：木雕工艺在中国具深厚悠久的历史，南派木雕工艺发展历经一千余年，沿承有序。其源可上溯到南宋首都临安（今杭州）建设时代，在南宋新都城的大规模营造中，徽州木匠是其中主要的技术工匠，当然随着木工匠人的到来，徽商同样给建造中的首都带来了无限商机，尤其是建造所需的木材和漆器。当这些徽商与徽州工匠致富后回到家乡，他们同样带回了新都最奢华的雕刻装饰和技艺，并将之用于徽州大量的村落民居建造之中。有着雄厚财力的徽商和纯熟技艺的工匠，使得雕刻（木雕、砖雕、石雕）在徽州地区很快流行起来并大量应用于宅邸，花园，公祠建筑之中。首先是木雕工艺飞速发展，然后砖雕和石雕也随之发展并形成徽派统一的风格。舞狮题材的木雕物件是徽州民间木雕技艺最杰出的代表，这些木雕狮子造型异常丰富，其造型与形态均具有南方狮子特有的憨态，调皮，且不失威严，这些狮子都是精心设计的。其动态造型与建筑木结构完美结合。既是木结构的重要支撑部件，也作为支架，月梁的重要装饰。徽州木雕的装饰纹样主要分为两类，一类为神话故事、民间传说、戏曲场景和神仙境界；另一类为真实生活中的人物鸟兽、吉祥花卉等。后者在普通民居建筑中更为普遍。

图 85：徽州民居月梁装饰：徽州民居中最大，最奢华的木雕装饰是厅堂中的梁柱，尤其是月梁上下的支架，托臂等工艺手法极为精湛。立柱的木雕装饰《刘海戏蟾图》，将传统民俗戏曲的场景直接用木雕艺术表现在建筑结构上，刘海戏蟾的故事在中国传承久远，即象征了科举成功，即所谓“蟾宫折桂”，也是对家庭幸福的祈愿。象征主人获得安宁、富足生活之寓意。

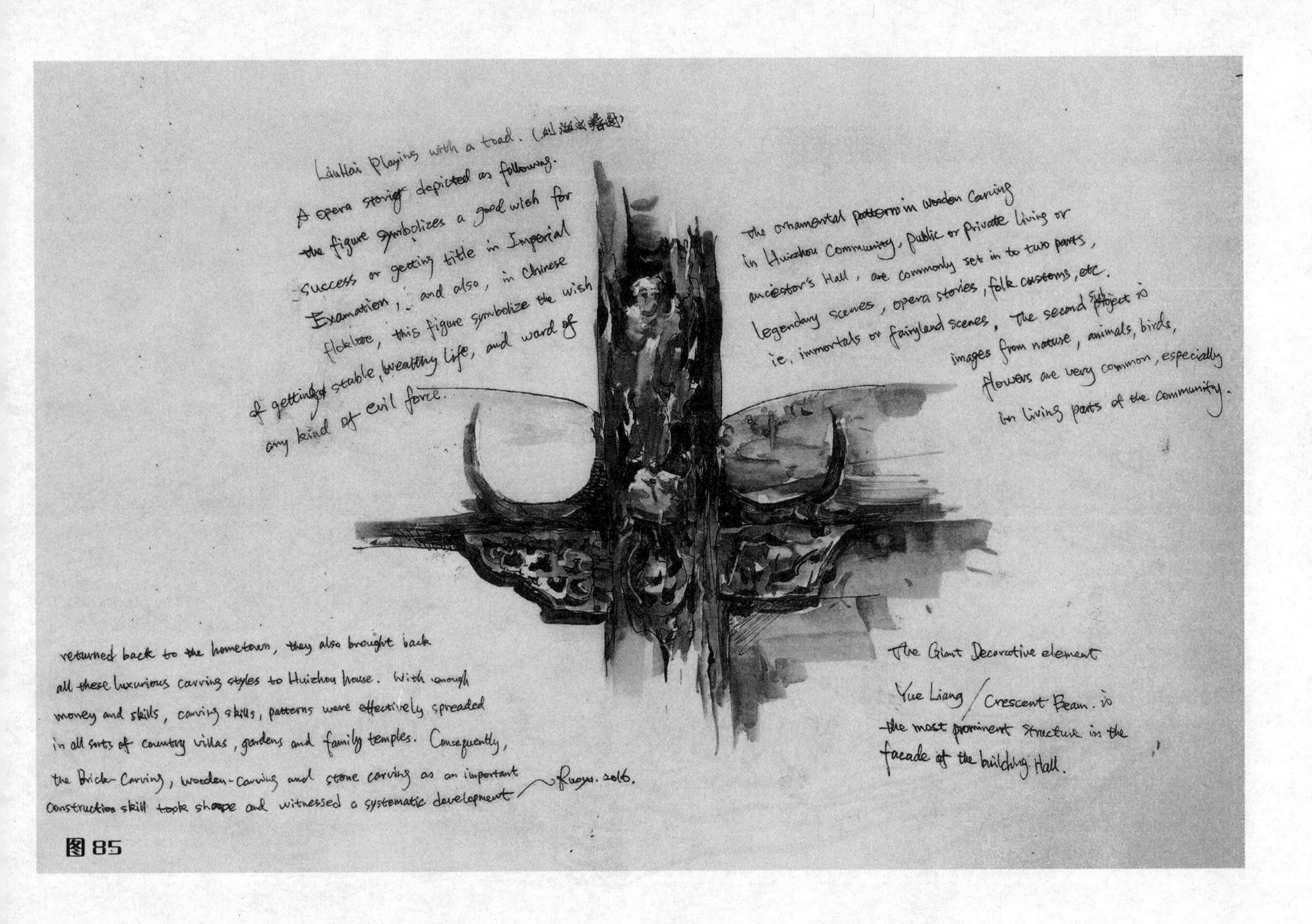
LiuHai Playing with a toad. (刘海戏蟾图)
A opera stories depicted as following.
the figure symbolizes a good wish for
success or getting title in Imperial
Examation, and also, in Chinese
flokloe, this figure symbolize the wish
of getting stable, wealthy life, and ward of
any kind of evil force.
The ornamental patterns in wooden carving
in Huizhou Community, public or private living or
ancestor's Hall, are commonly set in to two parts,
legendary scenes, opera stories, folk customs, etc.
ie, immortals or fairyland scenes, The second sub project is
images from nature, animals, birds,
flowers are very common, especially
in living parts of the community.
returned back to the hometown, they also brought back
all these luxurious carving styles to Huizhou house. With enough
money and skills, carving skills, patterns were effectively spreaded
in all sorts of country villas, gardens and family temples. Consequently,
the Brick-Carving, wooden-Carving and stone carving as an important
construction skill took shape and witnessed a systematic development
Ruoyu. 2016.
The Giant Decorative element
Yue Liang / Crescent Beam. is
the most prominent structure in the
facade of the building Hall.
图85

图86

Yunyang Hani Terraces, perticularly created by Chinese Hani ethnic group, is a perfect Combination of farming and housing civilization of South-West China, which developed for a long history of more than 1300 years. The iconic beauty of farming landscape created a perticular way of fine cultivation and high developed skills of irrigation of high lands, (a hanging garden or a heaven actually) which also embodied the unique tradition and Religious intention behind the glittering landscape.

As a property of landscape, it Comprises of the forest at the hilltop, the stockaded villages blow the forest, and the rice terraced fields just blow the villages, and the natural water system, the deep valley in most cases served as the bottom of the whole Landscape.

The terraces are the most outstanding earth art and Lanscape sculpture. For more than 1300 years, Hani people from generation to generation, continous cultivated this Land, carved out the greatest Land art and rural Landscape from natural Red Clay Mountains, every pieces of terrace twists with the terrain, winding like layers of waves, combines the most imposing Landscape, and a miracle of human farming civilization.

图 86：贵州的梯田：元阳哈尼梯田是中国西南地区悠久的农耕文明与聚落民居完美结合的产物。其发展历史超过 1300 余年。这种壮观的景色是气势恢宏的大地艺术与精耕细作的传统农业相结合后的产物，也是中国园林史上的“悬圃”（即空中花园）的真实再现，闪烁金光的田野映衬出的独特的居住方式和对大山的尊崇。作为一种传统的农业景观遗产，它是“山——水——林——田”的完美结合，最上层是山顶的林木（保持珍贵的表土不被雨水冲走），林木之下为聚落村寨；林木下便是连绵不绝的层层梯田，景观的最下层，接近山谷的位置是河湖水系，形成一幅层层向上，相互掩映的自然与人工完美融合的景观。

梯田景观是人类景观史上最杰出的大地雕刻作品，在超过千年的历史上，哈尼人代代相承，修田不绝。在红土地上开垦出千万沟壑和层层水田，每一层梯田都随着山坡走势婉转，形成极富魅力的动感线条与光色，创造出农耕文明的奇迹和大地艺术的杰作。

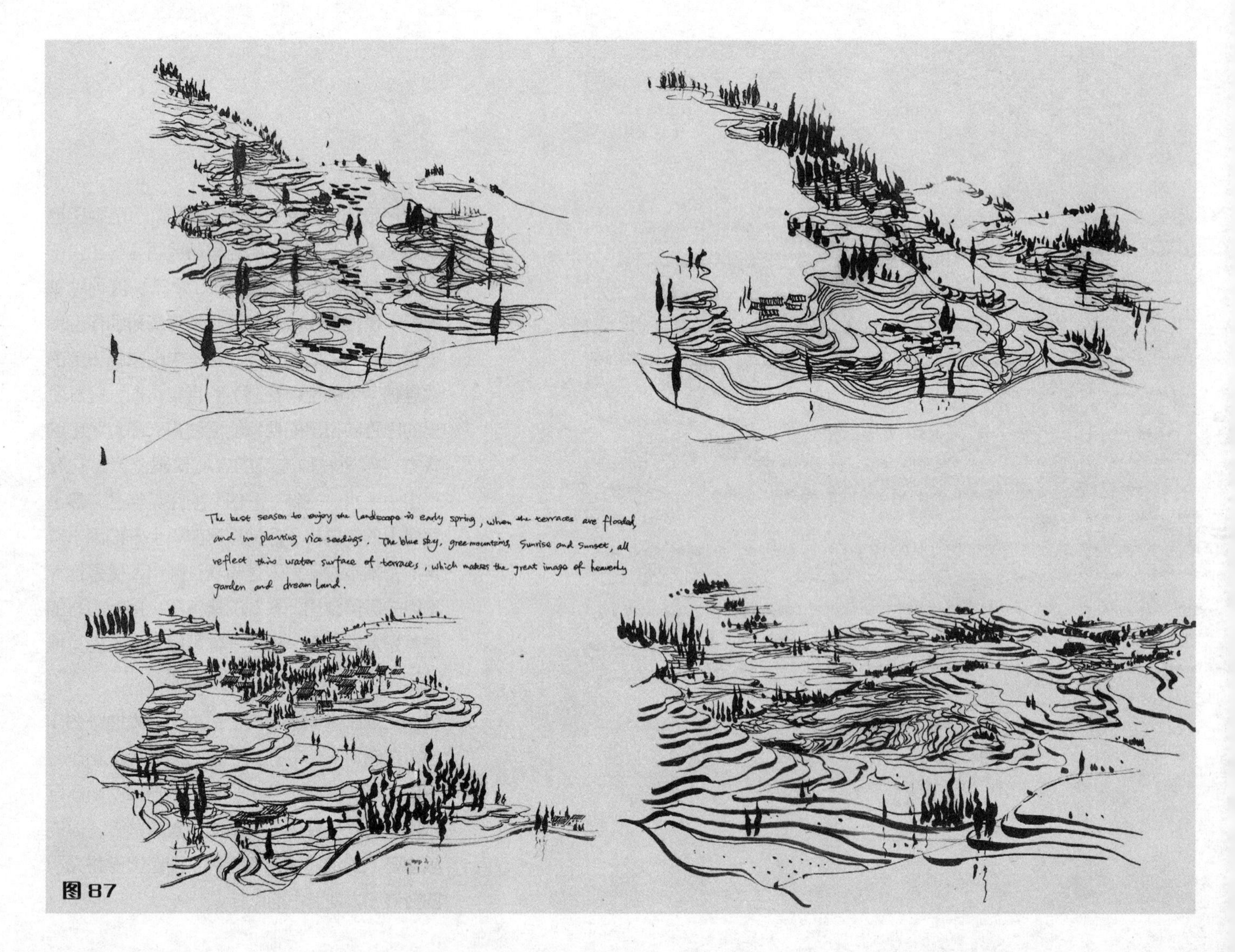
The best season to enjoy the landscape is early spring, when the terraces are flooded,
and we planting rice seedings. The blue sky, gree mountains, Sunrise and Sunset, all
reflect this water surface of terraces, which makes the great image of heavenly
garden and dream land.

图 87

Dancing Rhythm. 2017.7
To a large extent, the image of Yunnan Red Soil, always comes from these
gorgeous fields. From the tidy, well-constructed fields, anyone has to admit that
it is a human engineering feat only made by those hard working people. Without hard
work to maintain the terrace walls, and irrigation system, the only, precious topsoil would
wash down into the valley.

图 88

图 87：春季是欣赏最美风景的季节，那时候人们开始在汩汩水流的梯田里播撒水稻的种子，湛蓝的天空和绿色的山峦相互掩映，人们日出而作，日落而息，所有的场景都倒映在梯田的水面上，成为一幅幅田园风景画。

图 88：舞动的延伸——梯田的线条之美：人们对于红土地的深刻印象，很大程度上来源于这美丽的田园风景。这一层层精细耕作过的田野，是哈尼人世代勤恳劳作的见证，如果没有世世代代，不辞劳苦的人民进行修筑挡墙，输水上山，红土山上那仅有的一点点表土很快就会被降雨带入深谷之中，这些丰饶的水田也将不复存在。

A small mountain village
A small mountain village in the morning mist
Overview of a mountain village

图 89

图 90

A Barmyard at the foot of the hill. 2017.8. in Anlong.

图 91

图 89：贵州村落景观 / 贵州山村速写 / 晨雾中的小山村 / 山村鸟瞰

图 90：自然景观与聚落完美结合的人居环境，小山村坐落于山脚的缓坡之上，层层梯田和林市居于村落上的山中，景观的视线被层层抬升，一直延伸到远方的喀斯特岩石山顶。

图 91：山脚下的小农家院（贵州六盘水）

图 92

图 92：黄果树大瀑布

大瀑布
7. 若愚与老爸合作
於北京

Across Lu Lang to Bayi County in Winter
Alongside the Sichuan-Tibet Road, the journey was s
the Alpine Fir stubbornly grow on the snow-capped Moun

图 93

miles drive, we
the "Bayi Town"
before the big storm coming.
projects for Tibet,
generation for LuLang Mountain forest,
the Eco-security of Eastern-Tibet
, which is the place storing most of
and forests.
Rueye. 2016. 11.

图 93：在大雪纷飞的十一月，沿川藏公路翻越鲁朗林海所见到中国最大的高山冷杉原始森林，从 2000 多米至 4000 多米的高山上均有生长，主干高达 40 米以上，为中国仅存的未经大规模砍伐的原始森林，巨木良才居全国之冠。

经过 300 多千米的艰难行进，我们的车终于在大雪夜之前赶回了八一镇，这一地区作为青藏高原高寒区域与云贵高原人工地带的唯一缓冲区，保留着中国高山林区与高山草甸混合地带的完整生态格局。

图 94

图 94：西藏速写两幅，从林芝到八一镇路上的山景

贵州安龙国际工作营

LA Design Studio FOUR

2017 年 9 月，参与由中共安龙县委，安龙县人民政府，中国建筑学会史学分会建筑与文化学术委员会，亚洲城市与建筑联盟等单位主办的 2017AAUA 世界村落——第二届安龙可持续发展与设计国际夏令营。作为国际化协同发展设计在中国的重要设计研究项目，团队围绕贵州安顺地区特有的喀斯特地貌所形成的梯田农业景观，苗寨聚落景观等课题，结合生态设计要求，以安龙县的洞广、巧洞等村落为核心，进行了高效的团队协作设计和探讨。 作为团队中唯一的中学生，我与来自清华大学、哈佛大学、千叶大学的五名研究生共同承担了基础设施与景观设计的研究和创作。在英国 AA 建筑联盟学院的伊娃 · 卡斯特罗教授（Eva Castro Iraola）和美国普拉特学院李克 · 雷蒙教授（Enrique B. Limon Jr.）的指导下，完成了对设计场地的地形、水文、植被等景观生态方面的评价与分析，并以河岸生态景观恢复和改善设计为核心，完成了一系列有益的，可持续景观的设计研究和探讨。通过与研究生和设计师的协作，了解了与本专业相关的更多的职业特征、需求和更具前瞻性的理论和技术手段。以下为本次研习会的主要个人工作成果。

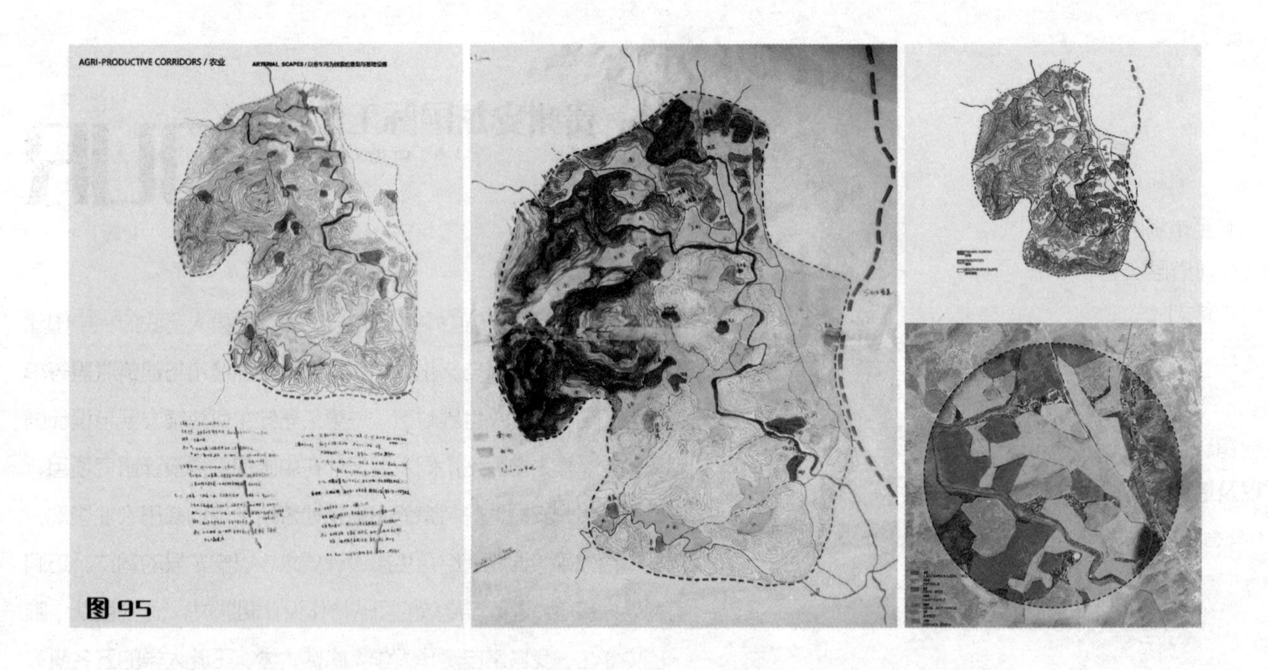

图 95、图 96：设计场地全部 12 个地块的总体评述

1-3 上者磨以西为连续的山坡谷地，西南向，光照充足，适合农业采摘等活动；高处有香车河支流，可引水灌溉。

缺点：交通不便。

建议：将该区域规划为二期开发建设：建成山地茶文化园，以茶为主题的民宿，工坊等。

4 下者磨南侧，两山之间的谷地，正南向，背山面水，引水灌溉，风景营造均较有利，适合建设香花谷地及贵州特色草药园（如安龙特色金银花园）、农业示范园。

5-6 上科喜——下科喜以北，为东南向谷地，坡度为 1 ：5，背西北主山，冬季面向东南暖坡。极有利于作为茶园，布依民俗园（大型茶山谷）

缺点：交通条件差，只有少量乡道连接。

建议：5-6 地块统一规则为农业茶文化、茶博会、药草园，并入二期建设。

7-8 中央花谷区：纳哈至木满的山谷，坡度适中（10%），景观廊道层次分明，处于核心山谷，交通较为便利（以桥与入口区直接相连），步行距离 500m。

建议：设中央花谷、百草园、体验园，两侧山坡设养生酒店民宿、禅修园等高端休闲场所，康练养生以及培训区。

9-10 贵落，鲁桃：山谷有大面积南向坡地，坡度大约 15%—25%，交通不便。

建议：建高端民宿，布依民俗文化村和体验馆等。

11-12 良台以北（接邻 313 省道），面积 30 万平方米，坡度 3%—10%，极平坦，北向和西北向为缓坡和平地，有利于大体量的公共设施建设，建议在此设立入口区，设置游客服务中心，演艺广场，露营中心和会议接待设施。其中 11 地块位于巧洞以北，面积 25 公顷，坡度 3%，为缓坡平地，建议设置入口标志景区以及香文化、茶文化和布依文化的综合展示区和购物街。

Site Analysis

N.1 and N.3 plot of land: Located in the west of Shangzhemo village faces the continuous hillside valley, southwest to the sunshine, is suitable for agricultura picking gardens and leisure activities;The high tributary of the Xiangche Rivei provides sufficient water source for irrigation.
Disadvantages: traffic inconvenience
Proposed projects: Tea Cultural Park, Tea Theme Homestay, Crafts Workshop, etc.

N.4 plot of land: Xiazhemo south, valley between two mountains, facing south,the back hill surface water, better to do the water diversion for irrigation and landscape construction, suitable for building herbs garden, fragrant flower valley and guizhou characteristics (such as Anlong characteristics of honeysuckle garden) anc agricultural demonstration garden.

N.5 and N.6 plot of land: The north of Shangkexi and Xiakexi facing southeast valley,or the winter warm slope (1:5),is the suitable place for tea garden and Buyi folk custom garden .The traffic conditions are poor, with only a few rural road connections.
Proposed projects: Tea culture and Tea Fair, herb garden, and reduce to the secor phase construction

N.7 and N.8 plot of land is the central flower valley. The valley stretching from Naha to Muman settlement lies in the core valley,with a moderate slope (10%), and a distinctive landscape corridor for central flower valley. The valley is connected directly to the main entrance by bridge 2 within a walking distance of 500m, is more convenient in the central area.
Proposed projects: central flower valley,herb gardens, experienced gardens, Zer garden, Health accommodation and homestay;

N.9 and N.10 plot of land located in Guiluo and Lutao settlement: The South valley is gently sloped terraces, with slope of 15% to 20%, the traffic is inconvenient,poor accessibility.
Proposed projects: High-end residential and Buyi folk culture village, experience museum, etc

N.11 and N.12 plot of land:It is a gentle slope of 30hm with average slope of 3% tc 10%,located in **north of Liangtai settlement,close to the 313 provincial road. It is a suitable place for big scale of** public facilities, like camping,performing arts plaza and receptior facilities like visiting center.
Proposed projects: Landmark for Main Entry; The Comprehensive Exhibition Area fo tea and Buyi culture; Shopping Street

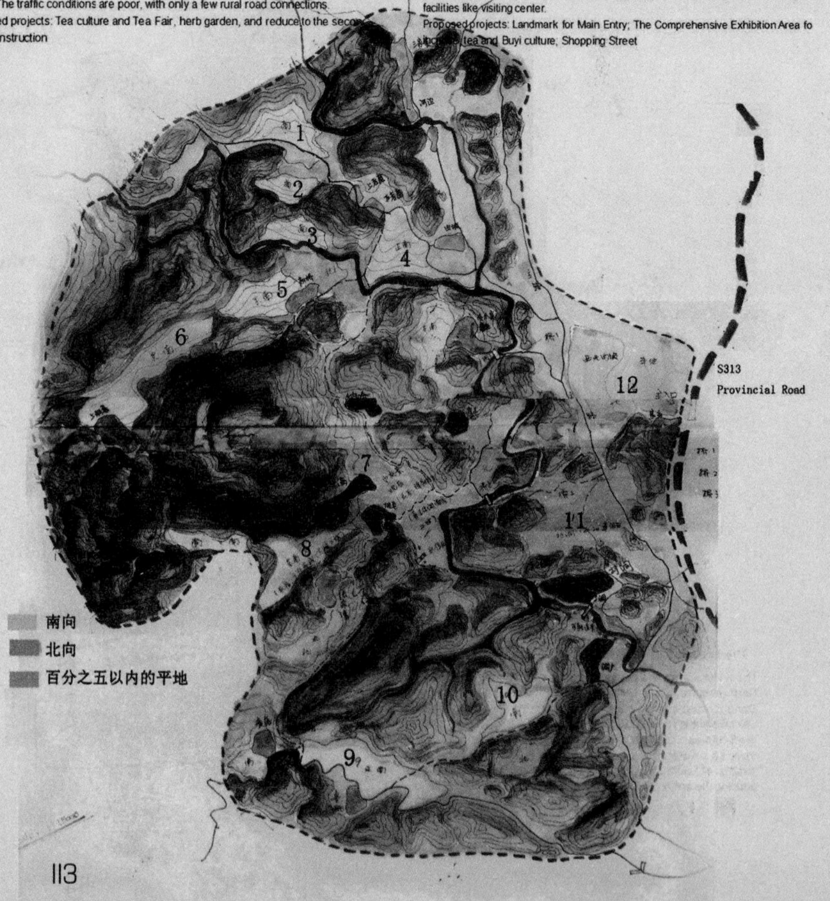

1
2
3
4
5
6
7
8
9
10
11
12
S313
Provincial Road
南向
北向
百分之五以内的平地

图 96

Feng Shui of Beijing

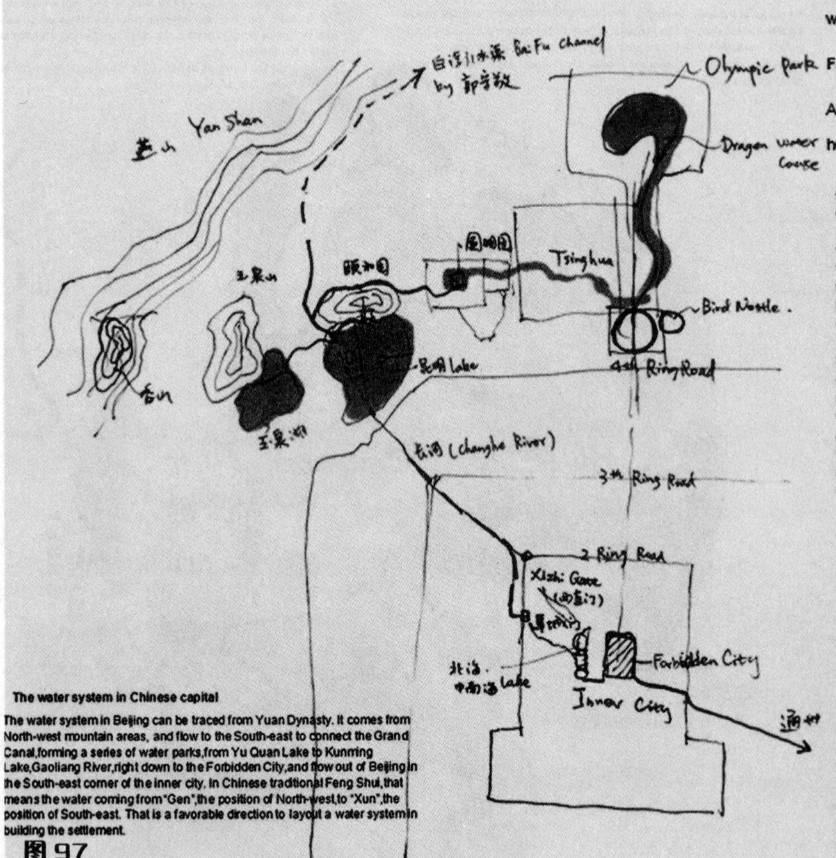

The water system in Chinese capital

The water system in Beijing can be traced from Yuan Dynasty. It comes from North-west mountain areas, and flow to the South-east to connect the Grand Canal,forming a series of water parks,from Yu Quan Lake to Kunming Lake,Gaoliang River,right down to the Forbidden City,and flow out of Beijing in the South-east corner of the inner city. In Chinese traditional Feng Shui,that means the water coming from "Gen",the position of North-west,to "Xun",the position of South-east. That is a favorable direction to layout a water system in building the settlement.

图 97

High Adaptable Waterbank

A elastic and multi-leveled bank system, make the riverbank more flexible to different water level, to suit the seasonal change of river.

Free spaces along the waterfront

A multi-layered riverbank leaves more room for different plantation, as well as various human activities, thus, creates more opportunities for human use.

C: The two village settlement all chose the high slope rather than lower valley as the site fo development; they all have good drainage conditions.

The ground elevation of each part of the settlement is fine controlled within the scope of the water wheel head; The layout of water channel in the village was decided in accordance wit the terrace landform configuration, in order to make full use of natural divide of landform, thus to achieve the non powered natural water circulation system. This water planning ides benefited both water supply and drainage for settlements, which shows a great wisdom o Chinese minority village in building the sustainable water environment.

D: Through the ingenious use of terrain landform, make the formation of mountair -settlement-river valley-river course,

settlement buildings smartly avoided the flood area and mountain rain runoff, and erosion as well.(as shown in figure)

At the same time, from the viewpoint of landscape, the gentle rising of river valley ,terracec field, and rolling slope makes an visual order for a multi-layered system, while the trail insertec in the terrace and alongside the water's edge, provides the best sight line lead to the Buy village and karst mountain peaks afar. (as shown in figure)

A Transfer from nature to human.

Elements::natural water system, artificial water system,irrigation system,dam,water gate, overflow dam..The most creative element:channel system, an exquisite man-made system that perfectly suit for the terraced field.(the irrigation canal connect with terrace system perfectly)----display the long-term tradition in Southern Chinese rural areas, a perfect combination of the needs for irrigation and settlement.

B. A self-powered water system

The water system in the village purely used the gravity rather than pumping process.(极少使用动力泵)the water was delivered step by step, from high to lower terraces. The inlet is always set at high branch while the outlet is at the lower one. The water level is controlled by water gate in storm and flooding time.

most creative element:channel system, an exquisite man-made system that perfectly sui for the terraced field.(the irrigation canal connect with terrace systerr perfectly)----display the long-term tradition in Southern Chinese rural areas, a perfec combination of the needs for irrigation and settlement.

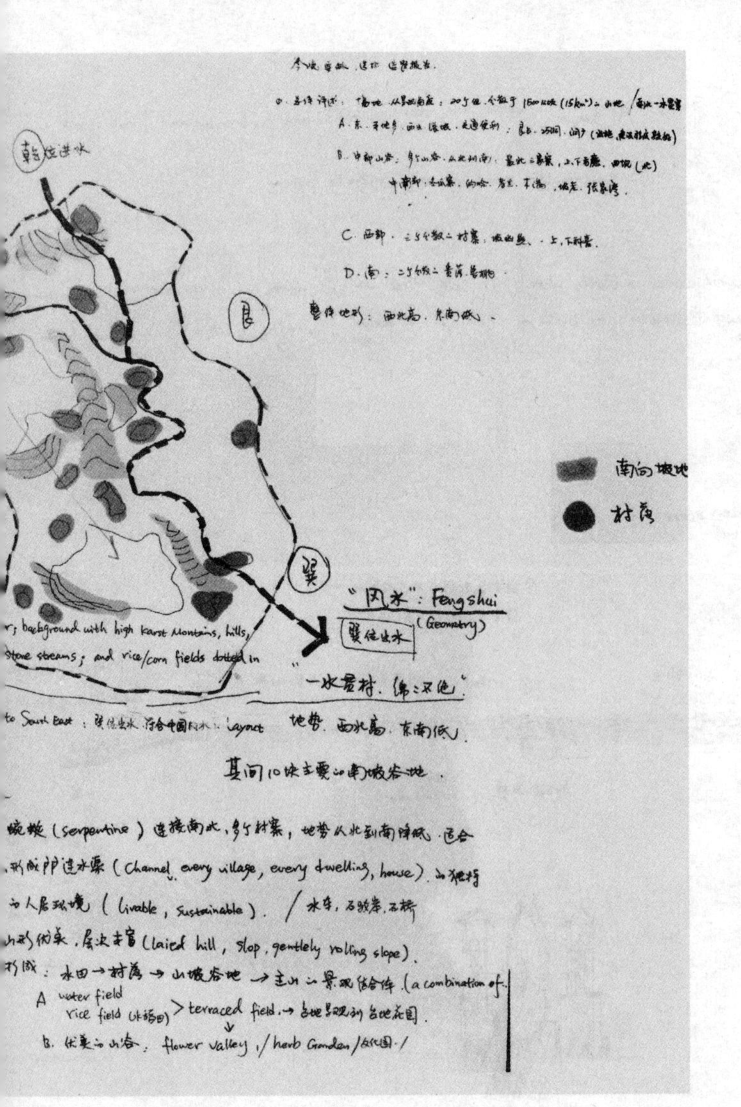

图 97：一水贯村，绵绵不绝

蜿蜒的水系将多个村寨连接在一起。背靠高高的喀斯特山脉，多条小河、小溪穿行其间。稻田和玉米田将整幅画卷点缀得更加美丽。

总结：

1. 水：水系蜿蜒，连接南北多个村寨，地势从北到南降低，适合引水，形成户户连水渠独特优越的人居环境（水车，石驳岸，石桥）。

水打造了一个宜居的人居环境。在该案例中，石桥、溪流为当地人提供了基本的生活条件。在香车河村，水最重要的特点在于其灵活性和流向。自如而又蜿蜒的水流自西北一路穿过整个村庄，流向东南方向。水系在当地人的生活中扮演着至关重要的角色。

中国的大江大河都是在这种地貌中诞生的，比如长江、黄河，它们都发源于等崇山峻岭之中，一路朝着东南方向奔流入海。这样的地貌和水系为中国营造了极为有利的生活环境。

2. 山：这里的居民可以完全依赖其背靠的山体。在我们的观念中，山具有生产性特征。所谓生产性是指其可以提供给人们赖以生存的资源，诸如树木、木材、食物、遮蔽物等。

香车河村山形优美，层次丰富，形成山水环抱的景观结构，以及从水田——村落——山谷坡地——喀斯特山峰的景观结合体。

A. 水田（稻田）——梯田——梯台式花园。

B. 美丽的山谷、花园、药草园——布依族文化园（少数民族文化园）

- A delicate way of fetching water to meet the flat transition from natural water system to artificial water supply system.

2. • A smart, water use that combines both production and living needs.

3. • A sustainable water use and recycle system.

巧洞村的水車設計.

Natural water → Water wheels → Self-flow → small Dam water → water level control (water Gate ...

Seasonal Release or Drain off → irrigation channel (水田.梯田) → Return back to t...

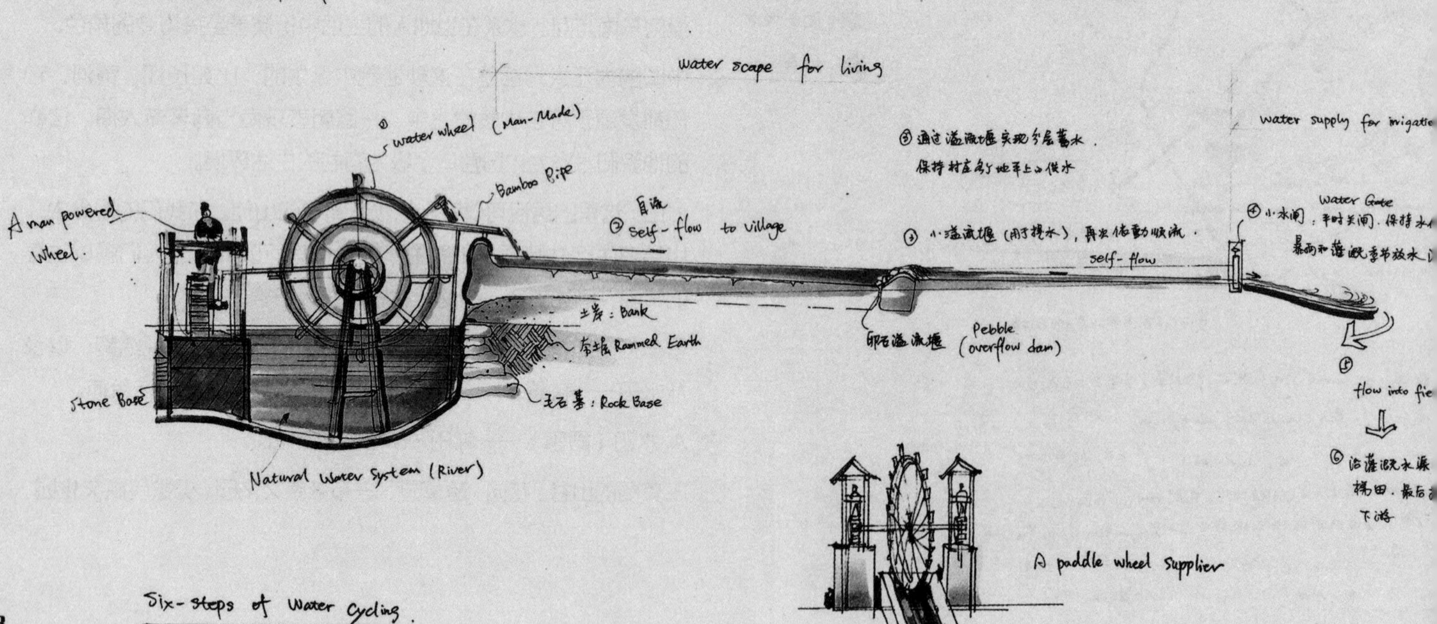

图 98

图 98：村落的水系统

香车河的水系统

富有创意的水车和村落水系统展示了华南乡野地区悠久的传统文明。它完美结合了灌溉需求和居住需求。

I. 理念：水系统是展现少数民族居住环境的重要元素。

2. 核心理念：将自然元素转变成人居环境。

元素：自然水系统、人工水系统、灌溉系统、水坝、水闸、溢流坝。

A . 最具创造力的水力设施和水渠，动力系统构造：

水利设施是一种精美的、极具创造力的人类工业文明产物，其非常完美地满足了梯田的灌溉需求。它也展示了华南乡野地区的一个传统文明，完美结合了灌溉需求和居住需求。

B. 一个完全依靠重力形成的水系自流系统构造：

自流水系统

村寨的水系统运转完全依靠重力，而非动力泵。水从高处到低处依次灌溉梯田。水系统的入口总是设在高处，出口设在低处。在下暴雨或者暴发洪水的时候，使用水闸来控制水位。

水岸系统：

高度适应性岸线设计——灵活的多级岸线系统使河堤可以适应不同的水位，以应对河流的季节变化。

自由的水岸空间——多层级的河堤可以适应不同作物的种植需求以及不同的人类活动。

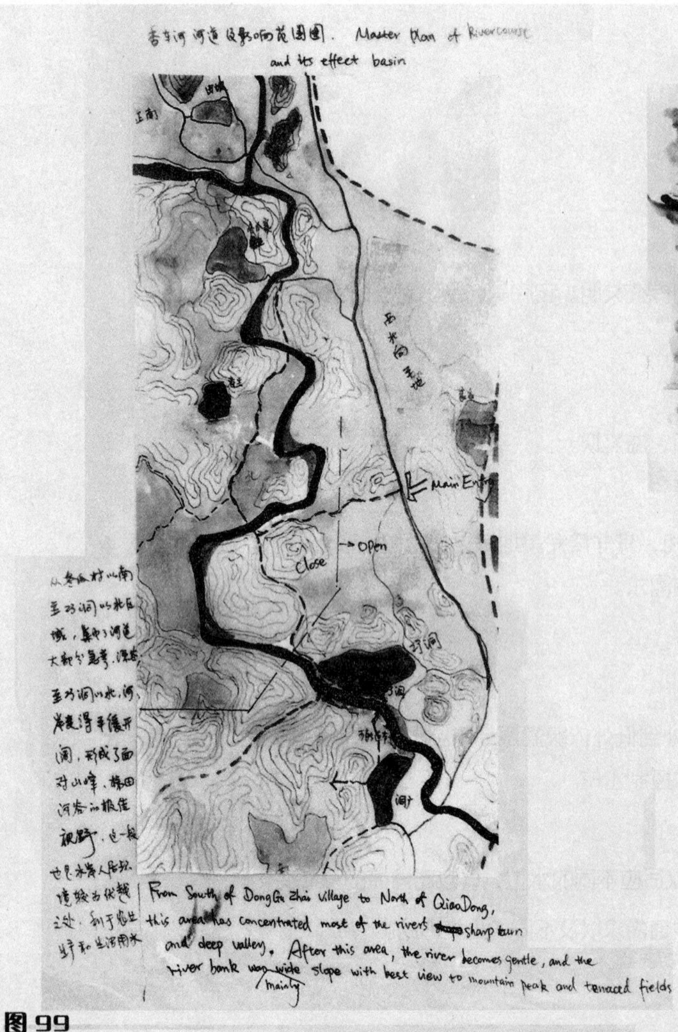

Master Plan of Rivercourse
and its effect basin
Main Entry
Close
Open
From South of DongGu Zhai Village to North of QiaoDong,
this area has concentrated most of the rivers sharp turn
and deep valley. After this area, the river becomes gentle, and the
river bank mainly wide slope with best view to mountain peak and terraced fields

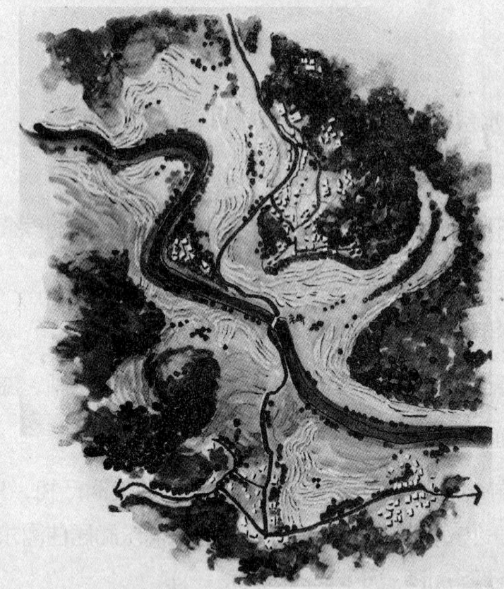

永久性河岸边界

村民的住宅，村落聚落建筑边界位于河岸之上的半坡高地和山顶之间

由于高度适中，可永久地避免水患，且取水便利；同时村中的引水及

排水渠也可充分利用自然落差形成多级供水，排水系统（Drainage）

《管子》：战国著名政治家，水利专家管子在其治国名著《管子》中（《管子·乘马》）

对这种中国古代聚落的落位方式进行过精辟的论述：

Sketch map of Dyna

Three Main Bankment

图 99

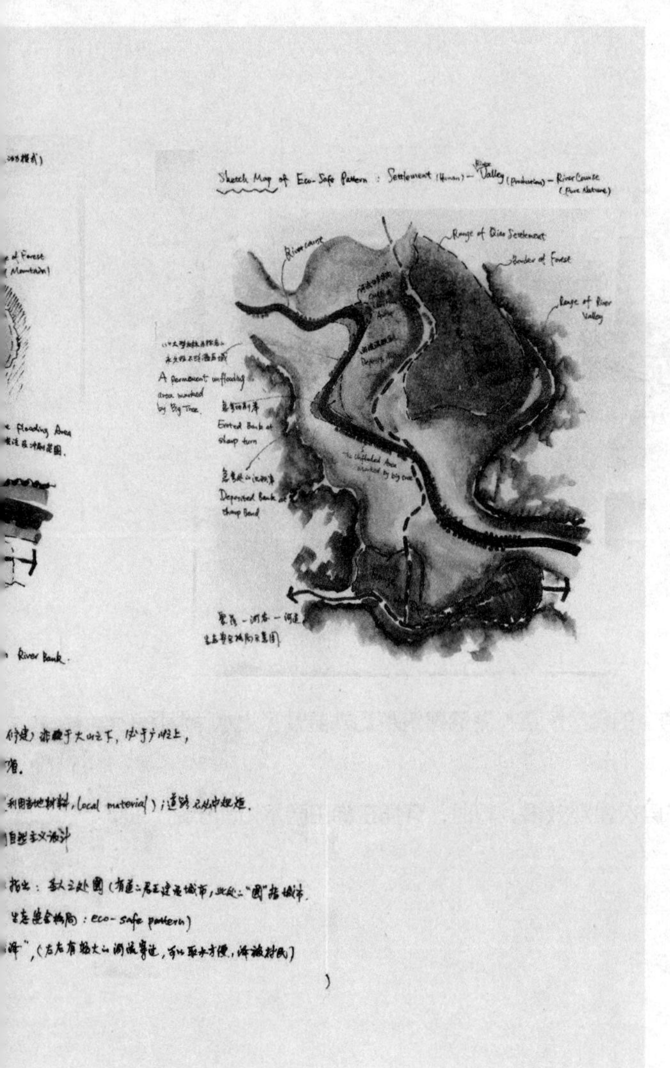

图 99: 香车河河道研究及巧洞——洞广段聚落景观生态格局分析

A: 两村均位于山脚与河谷高坡之间，取水便利且没有水患，属于永久性非淹没区；河谷的低处靠近河道，被留作梯田（生产用地），利于农田灌溉。

两村之间以主路隔开，该道路非常方便到达 313 省道。

B：村落住宅均朝向西南，这是贵州地区特有的落位方式。贵州山区光照不足，建筑群落多取南向偏西，以争取更多日照，解决村落民居采光的问题。巧洞——洞广——良台三个聚落之间的大部分场地平坦（坡度 5% 以内），朝向西南处的地方，是整个香车河村较大规模开发休闲文化的旅游用地。

本组在进行村落地形整体规划时，放弃一部分平坦的谷底地块，选择 10 块南向缓坡作为主要用地区域，并且大部分选择正南和西南两个方向。

C 两村聚落开发的位置均选择了高度适中的坡地，均有良好的排水条件，聚落各部分场地高差控制在水车扬程范围内（扬程即水车提水能抬升的最大高度），村内供水渠依照梯级配置，充分利用自然落差形成无动力的自然水循环系统，有利于河道供水和雨水排放。显示了中国少数民族村落水环境营造方面的高超智慧。

D：村落与河道的关系。

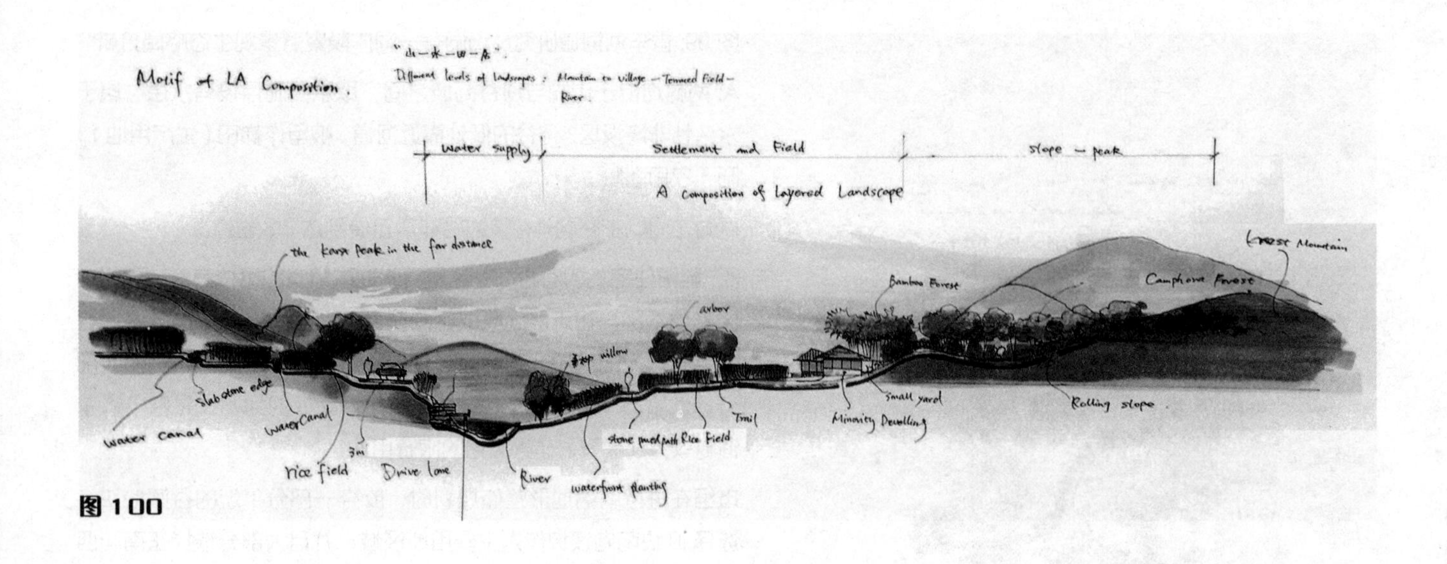

图 100

图 100：通过巧妙利用地形布局，形成了“山地——聚落——河谷——河道”的梯级配置，聚落建筑群巧妙避开了洪水泛滥区、山地雨水径流和冲刷区域。

同时，从景观角度看，河道、河谷、梯田的逐步提高，形成顺序展开的多层次景观效果。同时，穿插在梯田和水边的小路，提供了通向布依村寨和喀斯特山峰的最佳视觉廊道。

图 101：洞广村口竹桥堰坝设计

图 101

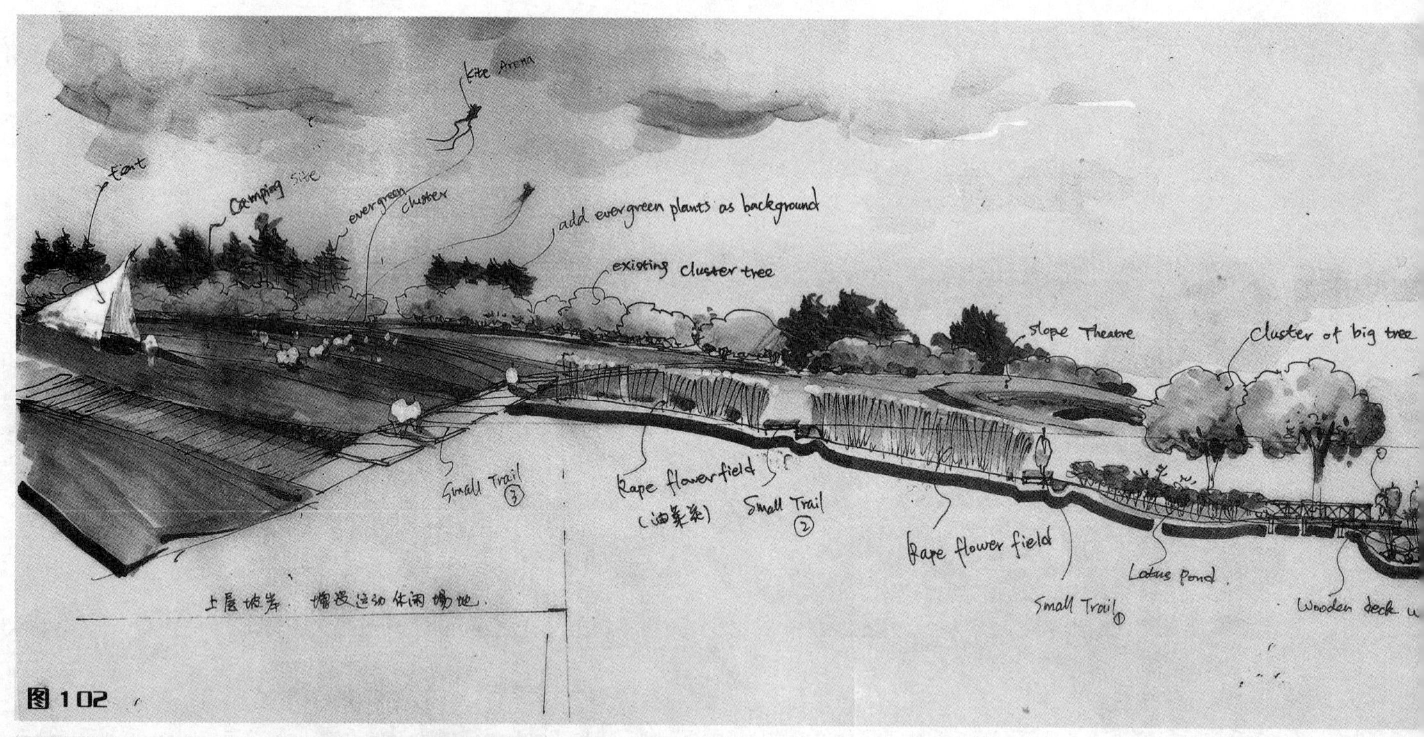
Kite Arena
tent
Camping Site
evergreen cluster
add evergreen plants as background
existing cluster tree
Slope Theatre
Cluster of big tree
Small Trail ③
Rape flower field
(油菜花)
Small Trail ②
Rape flower field
Lotus pond
Small Trail ①
Wooden deck
上层坡岸 增设运动休闲场地
图 102

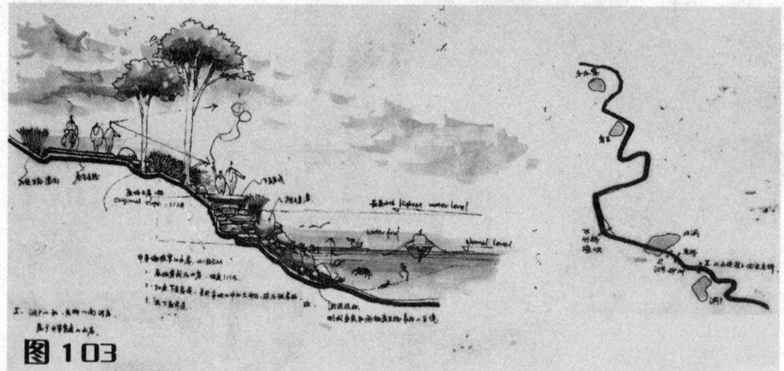
图 103

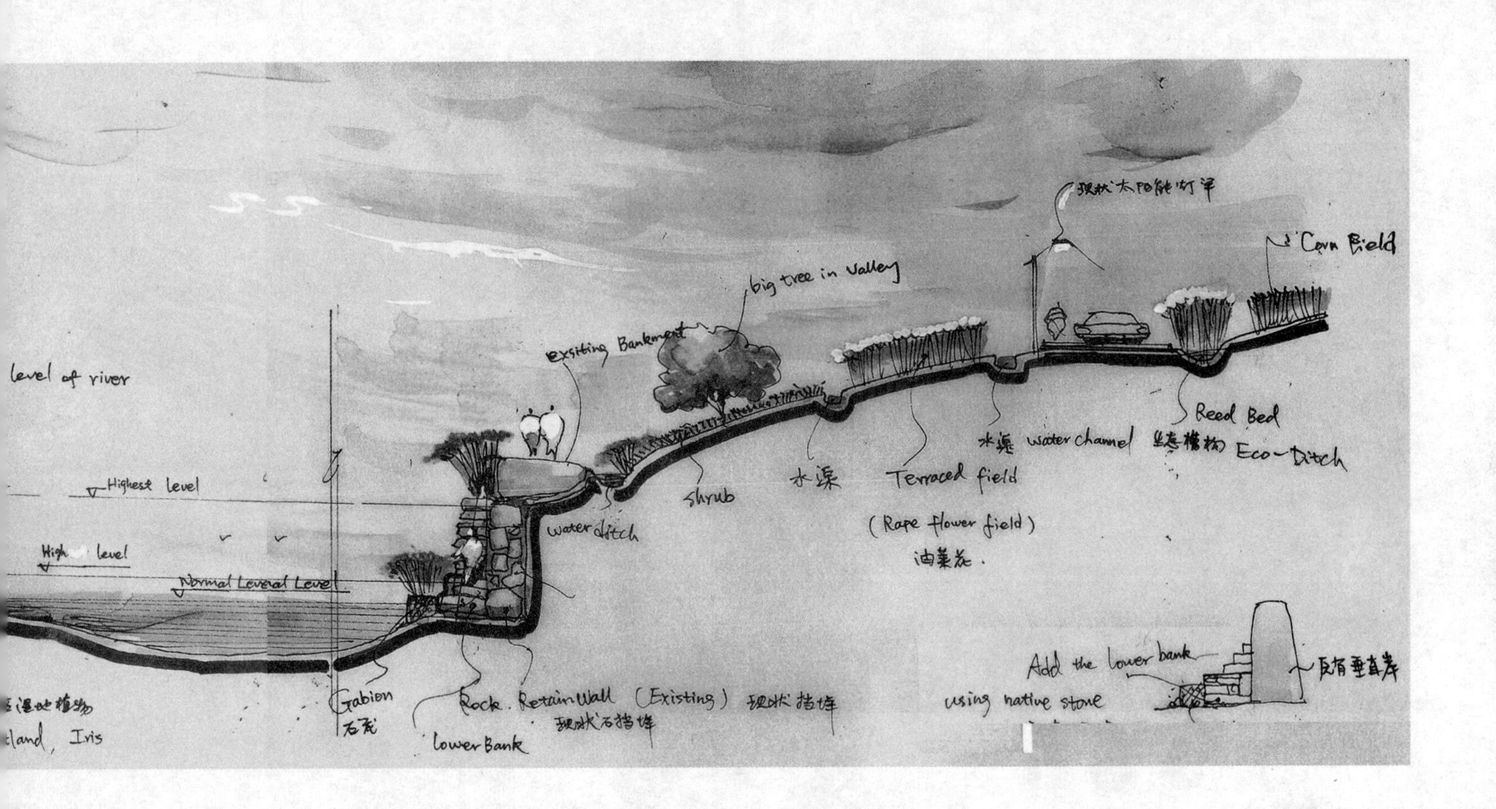

图 102：土方工程实现了多层水岸设计，可以满足不同生境和各种休闲活动要求。

使用各种乡土植被和乡土材料（木材、竹子、土、石头构筑物）等营造水岸景观。

图 103：使用乡土材料建造中等规模的河岸。

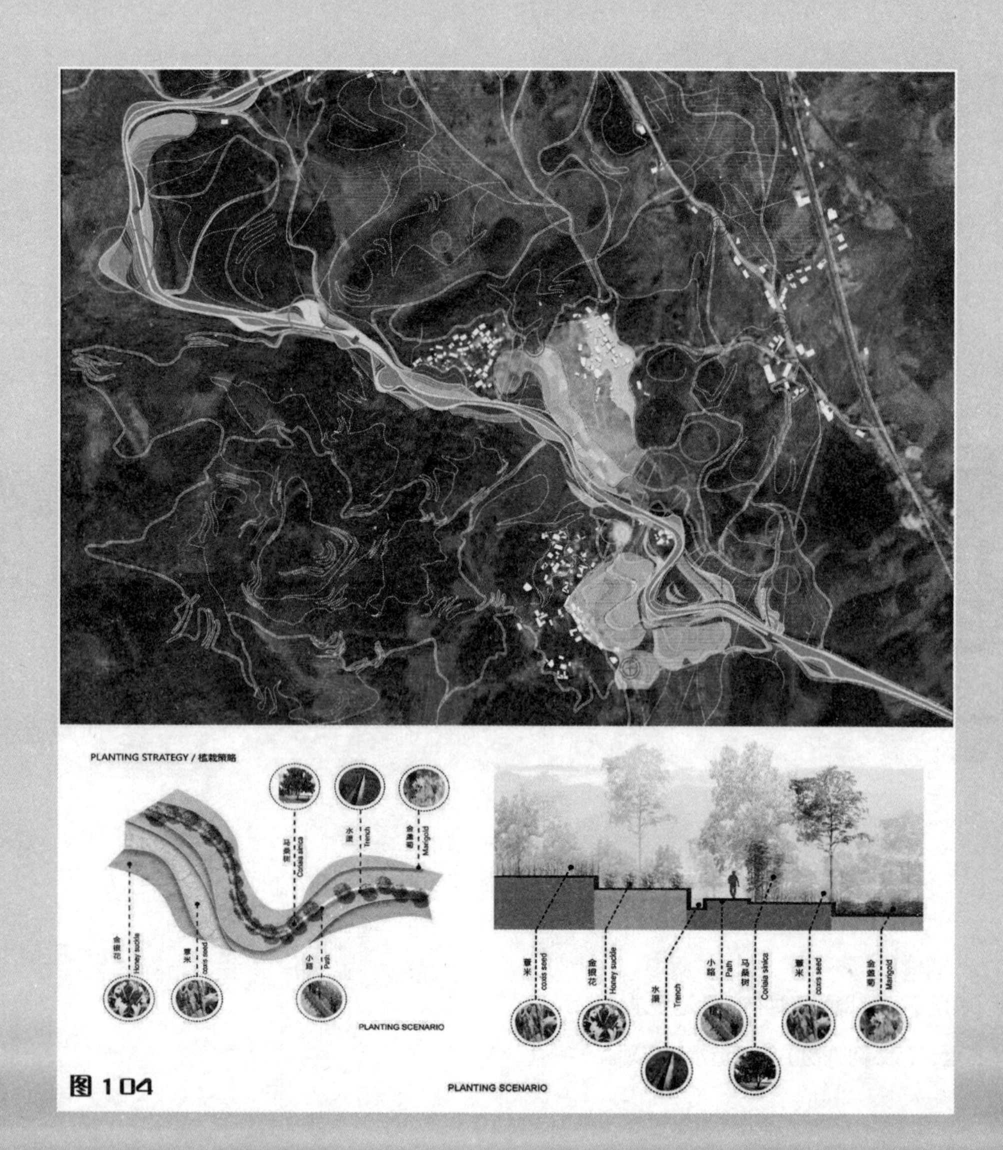
PLANTING STRATEGY / 植栽策略
Coriaia sinica
Trench
Marigold
Honey suckle
coxis seed
Path
PLANTING SCENARIO
coxis seed
Honey suckle
Trench
Path
Coriaia sinica
coxis seed
Marigold
PLANTING SCENARIO

图 104

图 104：植栽策略与种植场景

图 105：农场平面

图 106：洞广村口的多层次田园景观

图 107：巧洞村口广场的公共空间设计

图 107

图 108

图 108、图 109：安龙的农田景观设计草图

图 109

图 110

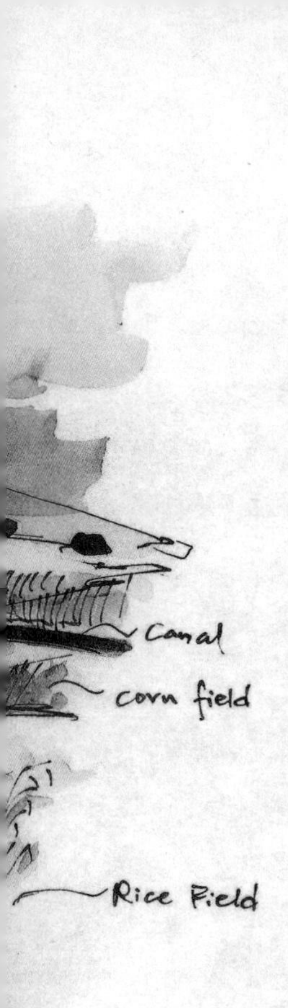

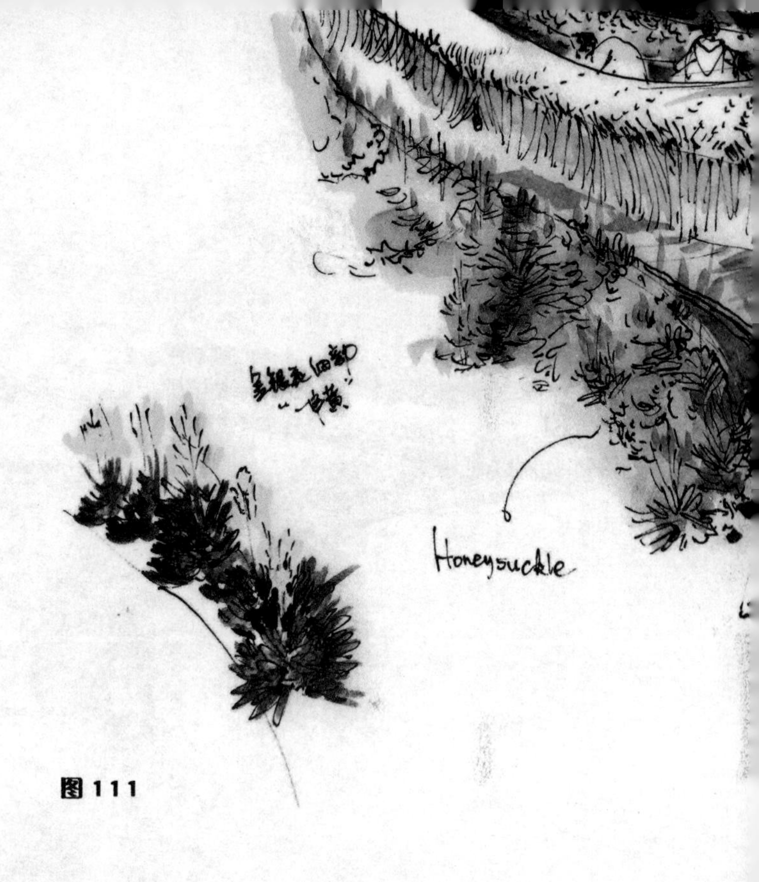

图 111

图 110、图 111：三种田地种植水稻，高粱，金银花

在现有的农田种植基础之上，突出贵州山区典型的农作物品种。形成水稻、高粱、金银花交替种植的大地图案。兼具美和生产效用的大地镶嵌景观。同时利用现有的农田水渠进行典型化处理，形成“水渠——农田——乡道——山坡林地”结合的大地镶嵌体。

图 112

图 113

图 112：三种田地种植
水稻，高粱，金银花

图 113：香车河河道陡
坡的河岸断面

图 114

图 114：贵州稻田速写三幅

图 115：（左）乡村小径 /（右）安龙县乡村的水田

Countryside Scenery
2017.7
in Anlong.
Rice, Field Landscape
in Anlong County.
2017.7

速写手卷 FIVE

Scroll-Design Section And Rendering

图116

图117

图 116：山坡上种着一株巨大的橡树。起伏的山坡上牛、羊成群遮蔽在橡树的阴影下，远处是欧洲红松。

图 117：温带林木与亚热带雨林比较——那不勒斯沿海的石松林。郁郁葱葱的苏格兰红松是开放的草坪、白色的宫殿和纯净的自然之间完美的中间元素，也是“英国花园中点缀的棕色”的独特地标。

图 118：一条绿带嵌在这个城市里，和大都会博物馆大片的绿色植物共同构成了曼哈顿的绿色中心。

图 118

图 119

图 119、图 120: 北师大校园

图 121、图 122：动物园剖面练习（亚洲熊园）

图 123

TENSIONING FOR MODERN AVIARY. Ruoyu. 2017. BeiJing

图 123：将古罗马贵族瓦罗式的鸟笼放大到现代动物园——当代鸟笼的张拉结构

图 124、图 125、图 126：动物园剖面练习（亚洲熊园）

图 124

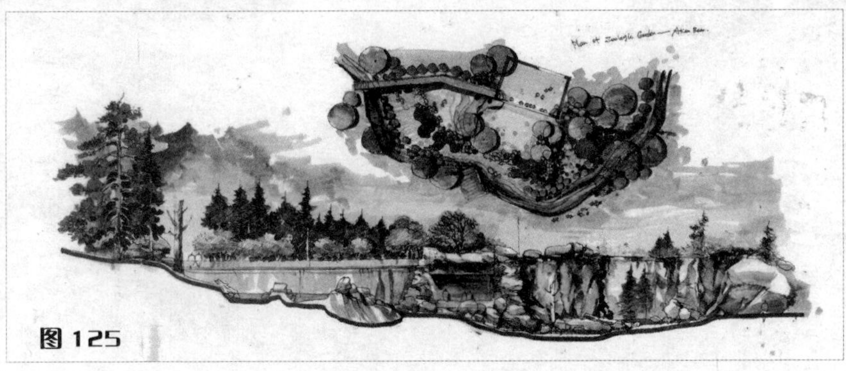

图 125

图 126

Actually, there are no big differences between a plate and a real
Round Ground, what ever a theater or just a simple clearing
in a forest. The only thing is were a different scale, besides
there are almost everything in common, So, the Best
50m
50m
way to be a good perspective drawing is to depict more things
like a plate, even though it's by no means simple, but it's clear,
pure. So I spent almost a whole week to practice on these things,
plates, tea cups, cooking pan, even wash basin in toilet.
I hope, next summer, if I don't take the Harvard Studio in Boston
I will surely return to
this office to
practice more about
these subjects, especially
those beautiful seafoods, and the Dream Burg.
Ruoyu. 2017.7 in Beijing,
University
Chili Sauce
this is Site

图 127

图 127: 事实上在画盘子和画圆形地面之间没有什么大的不同，无论是剧院还是森林里的一块阴影地。唯一的不同就是尺度规模。因此画好一个圆形场景透视的最好方式就是像画盘子一样，画更多类似盘子的东西，即便这样的透视画法也不简单，但它至少是清晰的，纯粹的。因此我花费了一周的时间去练习画盘子、茶杯、平底锅，甚至是卫生间里的洗脸盆。我希望，下个暑假，如果我没有去哈佛大学暑期国际工作营，我肯定会回到北京的工作室练习画这些物件，特别是很多漂亮的海鲜食物和汉堡包。

图 128：彩叶树练习—— 一片山林，一片秋色

图 129

图 129：场地练习

图 130：剖面练习 1：通过地形改造，将河岸分成上下两层，以适应滨水活动的不同需求。在冬天，河堤被市民当作自然滑雪场。

Ever Greenhedge
Eco-Ditch
Cherry Blossom
Greenway
Sideway
Reed Bed
Iris
By topographic Modification, the Riverbank is divided into upper and lower level, to adapt different Needs of Waterfront activities. While in winter, the Riverbank is used by citizens as a natural ski ground.

图 130

图 131、图 132：剖面练习 2

图 131

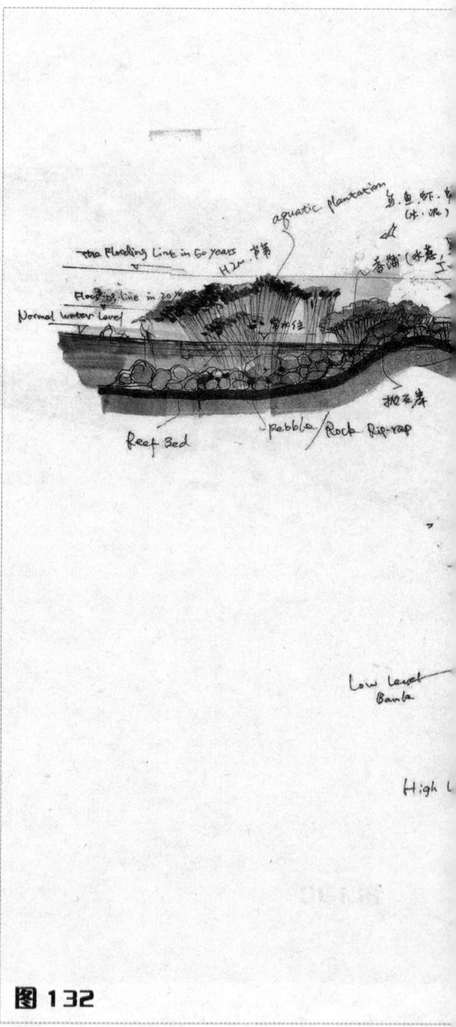

图 132

Upper Bank
Shrub
Camping
Green way for cycling
1:10
Iris
Bicycle Station of Greenway System
water front park
Spur Dike
Eco-Ditch
Greenway
cherry blossom
Plan of Pedestrain
Sideway
High way
overflow
Rainfall
rainfall
drainage
Ditch
The Bank of Spur Dike

图 133、图 134：剖面练习 3

图 133

图 134

愚者偶得 Scroll-Watercolor And Scribble SIX

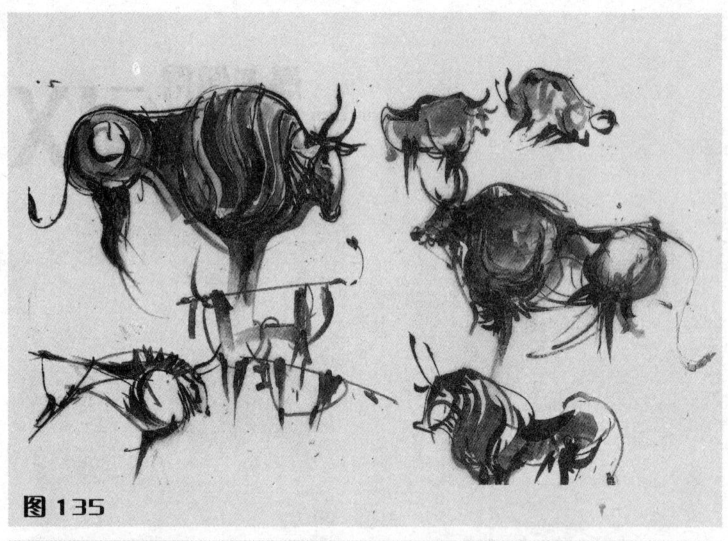

图 135

图 136

图 135、图 136、图 137：
绘画习作“乱涂乱画”

图 137

图 138

图 139

图 138、图 139、图 140：
绘画习作“乱涂乱画”

图 140

图 141、图 142、图 143：
绘画习作“乱涂乱画”

图 141

图 142

Double Bread
Onion
Double Cabbage
Chees
My Dream BigMac
onion
Cabbage
chees
Double patty
BigMac
cucumber
Beet slices
French Bagnette.
Chinese Chili Sauce.

图 143

图144

图145

图146

图144、图145、图146：绘画习作“乱涂乱画”

王若愚

王若愚，加拿大枫叶国际学校 II 年级学生，目前就读于温哥华昆特兰大学枫叶国际学校。从小学开始迷上风景速写和旅行考察，从小学三年级开始，作品相继获得北京海淀区小学生绘画比赛二等奖和全国特长生才艺竞赛三等奖等一系列奖项。循天性，秉家学，十年来，搜尽奇峰，勤练不辍，每遇心得辄手绘纪录，拣选成此风景情怀一书。或谓，愚者偶得，以飨同好学友。

图书在版编目（CIP）数据

风景的情怀 / 王若愚著. -- 南京 : 江苏凤凰文艺出版社, 2018.6
ISBN 978-7-5594-2179-1

Ⅰ. ①风… Ⅱ. ①王… Ⅲ. ①景观设计－建筑画－作品集－中国－现代 Ⅳ. ①TU204.132

中国版本图书馆CIP数据核字(2018)第106707号

书名	风景的情怀
著者	王若愚
责任编辑	孙金荣
特约编辑	杨琦 石慧勤
项目策划	凤凰空间／杨琦 石慧勤
出版发行	江苏凤凰文艺出版社
出版社地址	南京市中央路165号，邮编：210009
出版社网址	http://www.jswenyi.com
印刷	北京博海升彩色印刷有限公司
开本	710毫米×1000毫米 1/16
印张	10.25
字数	76 000
版次	2018年6月第1版
印次	2023年3月第2次印刷
标准书号	ISBN 978-7-5594-2179-1
定价	69.00元

（江苏凤凰文艺版图书凡印刷、装订错误可随时向承印厂调换）